Climatology: Concepts and Applications

Climatology: Concepts and Applications

Sophie Smith

STATES
ACADEMIC PRESS
www.statesacademicpress.com

States Academic Press,
109 South 5th Street,
Brooklyn, NY 11249, USA

Visit us on the World Wide Web at:
www.statesacademicpress.com

ISBN: 978-1-63989-108-5 (Hardback)

Cataloging-in-Publication Data

Climatology : concepts and applications / Sophie Smith.
p. cm.
Includes bibliographical references and index.
ISBN 978-1-63989-108-5
1. Climatology. 2. Climatic changes. 3. Meteorology. 4. Atmospheric science.
I. Smith, Sophie.

QC861.3 .C55 2022
551.5--dc23

Table of Contents

Preface

The scientific study of climate and weather is termed as climatology. It also attempts to study the causes of climate change and their impacts on human society. It is a subfield of physical geography which draws on the principles of biogeochemistry and oceanography. The principles of climatology are used for the analysis of accumulated data and pattern recognition. These patterns find applications in the field of weather forecasting. The subject of climatology can be categorized into various sub-disciplines such as paleoclimatology, historical climatology, boundary layer climatology and synoptic climatology. Some key areas of research in this field are climate variability, climate classification, climatological processes and climate change. This book is compiled in such a manner, that it will provide in-depth knowledge about the theory and practice of climatology. While understanding the long-term perspectives of the topics, it makes an effort in highlighting their impact as a modern tool for the growth of the discipline. In this book, constant effort has been made to make the understanding of the difficult concepts of climatology as easy and informative as possible, for the readers.

To facilitate a deeper understanding of the contents of this book a short introduction of every chapter is written below:

Chapter 1- Climatology is a branch of atmospheric science concerned with the description of climate and analysis of climatic differences, changes and their consequences. It helps us better understand the various atmospheric conditions that influence weather patterns and temperature changes over time. This chapter serves as an introduction to climatology and climate along with the methods associated with climate classification.

Chapter 2- The gaseous envelope surrounding the Earth is called the atmosphere. It is composed of a mix of several gases in varying amounts, with nitrogen being the most prominent. A study of its composition and structural layers is paramount in understanding atmospheric heating and cooling, insolation, atmospheric circulation, atmospheric moisture, etc. This chapter has been carefully written to provide an extensive understanding of the subject.

Chapter 3- Climatology can be divided into various branches; each corresponding to a specialized niche. Paleoclimatology, paleotempestology, historical climatology, tornado climatology, tropical cyclone rainfall climatology, urban climatology, dendroclimatology, etc. are a few branches of climatology examined in detail in this chapter.

Chapter 4- The ozone layer is a region located in the stratospheric layer of the Earth's atmosphere that absorbs much of the harmful ultraviolet rays emitted by the sun. It contains a high concentration of ozone gas which is instrumental in absorbing the ultraviolet rays. The release of CFCs and halons into the atmosphere results in chemical reactions which break down ozone molecules. The ozone Layer and its depletion are comprehensively covered in this chapter.

Chapter 5- Climate change is the global phenomenon of climate transformation characterized by long term changes in the weather patterns that have defined the Earth's local, regional and global climates. Activities like burning of fossil fuels and deforestation have contributed to this change. Melting of the polar ice caps, rising sea levels and global warming are some of the negative impacts of climate change. This chapter is dedicated to addressing this urgent issue that threatens the very survival of life on Earth.

Chapter 6- A climate model can be essentially defined as a representation of the climate system based on physical, biological and chemical principles. Climate models help in understanding the complex Earth systems and draw conclusions on past and future climate systems. General circulation model, mesothermal, microthermal, megathermal, etc. are a few of its aspects. This chapter has been carefully written to provide the reader with a better understanding of climate models.

Finally, I would like to thank the entire team involved in the inception of this book for their valuable time and contribution. This book would not have been possible without their efforts. I would also like to thank my friends and family for their constant support.

Sophie Smith

Chapter 1

Introduction to Climatology

Climatology is a branch of atmospheric science concerned with the description of climate and analysis of climatic differences, changes and their consequences. It helps us better understand the various atmospheric conditions that influence weather patterns and temperature changes over time. This chapter serves as an introduction to climatology and climate along with the methods associated with climate classification.

Climatology is the science of studying the average atmospheric conditions of a region, on long-term perspective. It includes a systematic study of the atmospheric conditions pertaining to both weather and climate. Climatology also aims to study the nature, causes and interpretations of its spatial variation and its association with the other spheres. It is closely related to both meteorology and geography. All the global atmospheric processes are spatially distributed and are varying with reference to time. It deals with all climatic components that are responsible in controlling these processes.

Climatic Components

The term Climate can be defined as the average weather conditions for a particular location and for a specific period of time. The average weather conditions are reflected in the form of climatologically parameters or variables. The very essential climatic parameters are temperature, precipitation, atmospheric pressure, humidity, and velocity and direction of winds. The elements of weather are recorded and analyzed by the scientists called climatologists. This is one of the oldest sciences of nature. And it is an ancient science, too.

The Ancient Science

These ancient literatures contain enormous information about the processes of cloud formation, occurrence of rain and the seasonal cycles that are caused by the movement of earth round the sun. Initially, the subject of meteorology was given much importance by the ancient Indians. This has happened as early as 3000 BC. There is a chronology behind the growth of climatological studies, in the world.

The Chronology

If we look back to the ancient observations, we may be able to find the core concepts and thoughts of this subject. It was around 600 BC, Thales described the water cycle. He was the first Greek meteorologist to work on this line. In 400 BC, Democritus predicted the changes in the weather with proper evidences. The first written records about the atmosphere were given by Hippocrates in the year 400 B.C. Aristotle writes about Meteorology, in the year 350 BC. Aristotle was the first to bring out the ever first compendium of Meteorology, as a weather science. It was mainly based on the works of the Greeks; the world was divided into three temperate zones. The tropical zone, temperate zone and the Frigid Zone are the basic divisions of the globe.

Climatology and Meteorology

The root for climatology lies in the subject of meteorology. It was the Greeks who tried to understand the atmosphere and its influence on life. The 20th century has witnessed a lot of advancements in the development of tools and techniques in order to obtain the regular and accurate data of weather and climate. Technological advancements have helped to operationalize many of the applications. Satellite based, real-time and continuous monitoring of all the segments of the whole globe, helps to understand the trends of weather and climate, well in advance.

The primary goal of Climatology is to study the unique characteristics of atmosphere, its structure and composition in controlling the global climate, origin and types of climates, causes and processes influencing the climatic variations, elements of weather and the impact of climate on human and other life. Agricultural climate, aviation climate, bio climate, medical climate, macro climate, micro climate, meso climate, Palaeo climate and Arctic climate are some areas of special studies coming under the broad science of climatology.

Climatology is an interdisciplinary subject. It involves the basic principles and application areas of physics, chemistry, geology, mathematics, electronics, botany, zoology and environmental science. It has direct links with the subjects like oceanography, forestry, hydrology, glaciology, agriculture, animal husbandry, fisheries, horticulture, floriculture, marine biology, microbiology. Today, climatology is a well grown field of science. It has several major and minor branches.

Major and Minor Branches

The major branches are:

- Historical Climatology,
- Physical and Dynamic Climatology,
- Regional Climatology,
- Satellite climatology,
- Tropical Climatology,
- Palaeoclimatology,
- Applied Climatology.

Diversified Fields

Historical climatology deals with the study of historical changes in climate and their effect on human history and development. The Physical Climatology is mainly concerned with the various atmospheric phenomena and their resultant climates. Regional Climatology deals with the regional variations of climatic conditions which include Micro climate, Local climate, Meso Climate and Macro Climate. Paleoclimatology is yet another branch which encompasses the effects of climatic change over the entire history of the earth. Applied Climatology deals with climatic controls of human activities and the application of climatic knowledge in solving unique problems faced by human population. In this respect, climatology is fully related to ecology.

Climatology and Ecology

Ecology is the study of the mutual relationship between organisms and their environment the study of earth's ecosystems involve both biotic and abiotic components. The environment of living organisms is directly affected by weather and climate, including those changes in climate that are gradually being made by human beings. Climatology has a direct bearing on ecology, ecosystems and the well-being of all life. They are all based on the global climate and cyclical processes.

Global Climate and Cyclical Processes

The global climate and many cyclical processes on the earth are all controlled by the atmosphere and its processes. These processes have both direct and indirect influence on several impacts on the world. Among those, the Weather and Climate are the two distinctive features that are having very direct influence on the whole globe and the life. The science of climatology draws full attention, due to these impacts on the earth and life. The four dimensions of the subject are Structure and dynamics of the atmosphere, Suns' radiation, global cyclic processes and their impacts on all segments of the earth. The first one is related to the atmosphere.

Climatology and Atmosphere

Atmosphere is an important segment of the globe, containing air, water vapour and particulate matter. Atmosphere is an envelope is attached with the globe due to the gravitational pull of the earth. The life on the Earth exists because of the presence of atmosphere. It protects all life by blocking out the some of the harmful radiation coming from the sun. Atmosphere is fully inter-linked with all other spheres like hydrosphere, lithosphere and biosphere. The weather and climate are a part of the atmospheric effects, which control the life and life support systems. The primary aspect in the subject of climatology is in analyzing the atmospheric variables and data acquisition. In order to do that, it is necessary to understand the Atmosphere, atmospheric chemistry, the Gas Laws, the Vertical Structure of The Earth's Atmosphere.

Importance of Atmosphere

Atmosphere is the multilayered gaseous envelope attached to the planet earth. It is a unique sphere of air, gases and water vapor. The height of this huge mass is estimated to range from 16000 to 29000 KM from the sea level. This is an essential environmental segment for the life support systems of biosphere and hydrosphere. The vertical variations in temperature and concentration of gases have a significant impact on global scales. The layer concerning with weather and climate, is troposphere. Atmosphere is composed of three major constituents as gases, water vapor and aerosols. Some are permanent (or) constant gases and some are variable gases. They obey Boyle's gas Law, Lussac's Law; Charle's Law and combined gas law.

The Study of Atmosphere

Under the umbrella of Climatology, the structure and composition of the atmosphere are studied first. An understanding of the thickness and disposition of different layers like troposphere, stratosphere and mesosphere are needed. Under the conditions of the atmosphere the following aspects are to be understood.

- Elements of Weather and Climate.
- Structure and Composition of the Atmosphere.
- Insolation and Energy balance.
- Atmospheric Temperature.
- Atmospheric Pressure and Motion.

Climatology and Global Energy

Solar energy and gravitational energy are the fundamental sources of energy for the Earth's climate system. The Sun is the source of the energy that causes the motion of the atmosphere and thereby controls weather and climate. Any change in the energy from the Sun received at the Earth's surface will therefore affect climate. During stable conditions there has to be a balance between the energy received from the Sun and the energy that the Earth radiates back into Space. This energy is mainly radiated in the form of long wave radiation corresponding to the mean temperature of the Earth.

Climatology Involves the Study of:

- Insolation,
- Flow of Energy,
- Absorption of radiation,
- Emission of Radiation,
- Distribution of Radiation,
- Heat Budget (or) Energy balance,
- Human influence on radiation balance.

Climate

"Climate" refers to the average weather in terms of the mean and its variability over a certain time-span and a certain area. Classical climatology provides a classification and description of the various climate regimes found on Earth. Climate varies from place to place, depending on latitude, distance to the sea, vegetation, presence or absence of mountains or other geographical factors.

Climate varies also in time; from season to season, year to year, decade to decade or on much longer time-scales, such as the Ice Ages. Statistically significant variations of the mean state of the climate or of its variability, typically persisting for decades or longer, are referred to as "climate change". The Glossary gives definitions of these important and central notions of "climate variability" and "climate change".

Climate variations and change, caused by external forcing, may be partly predictable, particularly on the larger, continental and global, spatial scales. Because human activities, such as the emission of greenhouse gases or land-use change, do result in external forcing, it is believed that the large-scale aspects of human-induced climate change are also partly predictable. However the ability to actually do so is limited because we cannot accurately predict population change, economic change, technological development, and other relevant characteristics of future human activity. In practice, therefore, one has to rely on carefully constructed scenarios of human behavior and determine climate projections on the basis of such scenarios.

Climate Variables

The traditional knowledge of weather and climate focuses on those variables that affect daily life most directly: average, maximum and minimum temperature, wind near the surface of the Earth, precipitation in its various forms, humidity, cloud type and amount, and solar radiation. These are the variables observed hourly by a large number of weather stations around the globe.

However this is only part of the reality that determines weather and climate. The growth, movement and decay of weather systems depend also on the vertical structure of the atmosphere, the influence of the underlying land and sea and many other factors not directly experienced by human beings. Climate is determined by the atmospheric circulation and by its interactions with the large scale ocean currents and the land with its features such as albedo, vegetation and soil moisture. The climate of the Earth as a whole depends on factors that influence the radiative balance, such as for example, the atmospheric composition, solar radiation or volcanic eruptions. To understand the climate of our planet Earth and its variations and to understand and possibly predict the changes of the climate brought about by human activities, one cannot ignore any of these many factors and components that determine the climate. We must understand the climate system, the complicated system consisting of various components, including the dynamics and composition of the atmosphere, the ocean, the ice and snow cover, the land surface and its features, the many mutual interactions between them, and the large variety of physical, chemical and biological processes taking place in and among these components. "Climate" in a wider sense refers to the state of the climate system as a whole, including a statistical description of its variations.

The Climate System

Its Components

The climate system is an interactive system consisting of five major components: the atmosphere, the hydrosphere, the cryosphere, the land surface and the biosphere, forced or influenced by various external forcing mechanisms, the most important of which is the Sun. Also the direct effect of human activities on the climate system is considered an external forcing. The atmosphere is the most unstable and rapidly changing part of the system. Its composition, which has changed with the evolution of the Earth, is of central.

The Earth's dry atmosphere is composed mainly of nitrogen (N_2, 78.1% volume mixing ratio), oxygen (O_2, 20.9% volume mixing ratio, and argon (Ar, 0.93% volume mixing ratio). These gases have only limited interaction with the incoming solar radiation and they do not interact with the infrared radiation emitted by the Earth. However there are a number of trace gases, such as carbon dioxide (CO_2), methane (CH_4), nitrous oxide (N_2O) and ozone (O_3), which do absorb and emit infrared radiation. These so called greenhouse gases, with a total volume mixing ratio in dry air of less than 0.1% by volume, play an essential role in the Earth's energy budget. Moreover the atmosphere contains water vapour (H_2O), which is also a natural greenhouse gas. Its volume mixing ratio is highly variable, but it is typically in the order of 1%. Because these greenhouse gases absorb the infrared radiation emitted by the Earth and emit infrared radiation up- and downward, they tend to raise the temperature near the Earth's surface. Water vapour, CO_2 and O_3 also absorb solar short-wave radiation.

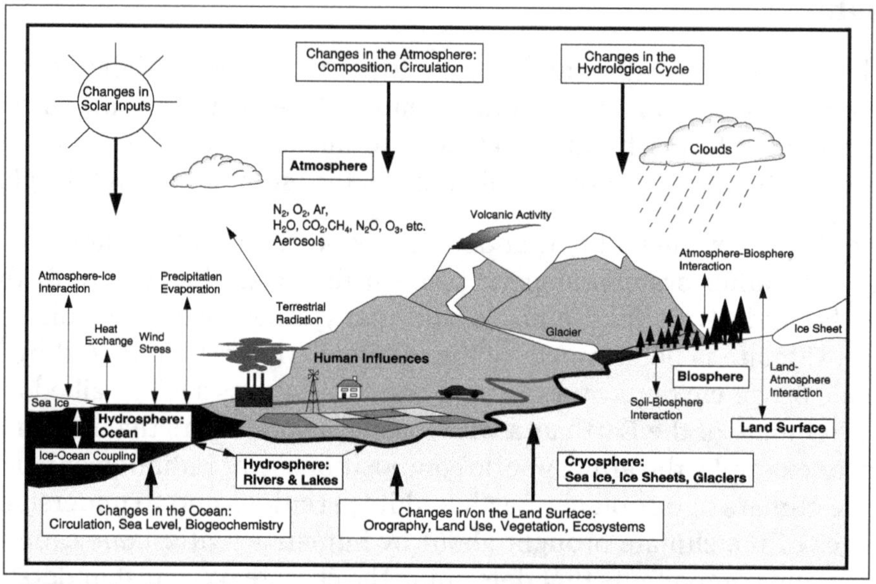

Figure: Schematic view of the components of the global climate system (bold), their processes and interactions (thin arrows) and some aspects that may change (bold arrows).

The atmospheric distribution of ozone and its role in the Earth's energy budget is unique. Ozone in the lower part of the atmosphere, the troposphere and lower stratosphere, acts as a greenhouse gas. Higher up in the stratosphere there is a natural layer of high ozone concentration, which absorbs solar ultra-violet radiation. In this way this so-called ozone layer plays an essential role in the stratosphere's radiative balance, at the same time filtering out this potentially damaging form of radiation.

Beside these gases, the atmosphere also contains solid and liquid particles (aerosols) and clouds, which interact with the incoming and outgoing radiation in a complex and spatially very variable manner. The most variable component of the atmosphere is water in its various phases such as vapour, cloud droplets, and ice crystals. Water vapour is the strongest greenhouse gas. For these reasons and because the transition between the various phases absorb and release much energy, water vapour is central to the climate and its variability and change.

The hydrosphere is the component comprising all liquid surface and subterranean water, both fresh water, including rivers, lakes and aquifers, and saline water of the oceans and seas. Fresh water runoff from the land returning to the oceans in rivers influences the ocean's composition and circulation. The oceans cover approximately 70% of the Earth's surface. They store and transport a large amount of energy and dissolve and store great quantities of carbon dioxide. Their circulation, driven by the wind and by density contrasts caused by salinity and thermal gradients (the so-called thermohaline circulation), is much slower than the atmospheric circulation. Mainly due to the large thermal inertia of the oceans, they damp vast and strong temperature changes and function as a regulator of the Earth's climate and as a source of natural climate variability, in particular on the longer time-scales.

The cryosphere, including the ice sheets of Greenland and Antarctica, continental glaciers and snow fields, sea ice and permafrost, derives its importance to the climate system from its high reflectivity (albedo) for solar radiation, its low thermal conductivity, its large thermal inertia and, especially, its critical role in driving deep ocean water circulation. Because the ice sheets store a large amount of water, variations in their volume are a potential source of sea level variations.

Vegetation and soils at the land surface control how energy received from the Sun is returned to the atmosphere. Some is returned as long-wave (infrared) radiation, heating the atmosphere as the land surface warms. Some serves to evaporate water, either in the soil or in the leaves of plants, bringing water back into the atmosphere. Because the evaporation of soil moisture requires energy, soil moisture has a strong influence on the surface temperature. The texture of the land surface (its roughness) influences the atmosphere dynamically as winds blow over the land's surface. Roughness is determined by both topography and vegetation. Wind also blows dust from the surface into the atmosphere, which interacts with the atmospheric radiation.

The marine and terrestrial biospheres have a major impact on the atmosphere's composition. The biota influences the uptake and release of greenhouse gases. Through the photosynthetic process, both marine and terrestrial plants (especially forests) store significant amounts of carbon from carbon dioxide. Thus, the biosphere plays a central role in the carbon cycle, as well as in the budgets of many other gases, such as methane and nitrous oxide. Other biospheric emissions are the so-called volatile organic compounds (VOC) which may have important effects on atmospheric chemistry, on aerosol formation and therefore on climate. Because the storage of carbon and the exchange of trace gases are influenced by climate, feedbacks between climate change and atmospheric concentrations of trace gases can occur. The influence of climate on the biosphere is preserved as fossils, tree rings, pollen and other records, so that much of what is known of past climates comes from such biotic indicators.

Interactions among the Components

Many physical, chemical and biological interaction processes occur among the various components of the climate system on a wide range of space and time scales, making the system extremely complex. Although the components of the climate system are very different in their composition, physical and chemical properties, structure and behaviour, they are all linked by fluxes of mass, heat and momentum: all subsystems are open and interrelated.

As an example, the atmosphere and the oceans are strongly coupled and exchange, among others, water vapour and heat through evaporation. This is part of the hydrological cycle and leads to condensation, cloud formation, precipitation and runoff, and supplies energy to weather systems. On the other hand, precipitation has an influence on salinity, its distribution and the thermohaline circulation. Atmosphere and oceans also exchange, among other gases, carbon dioxide, maintaining a balance by dissolving it in cold polar water which sinks into the deep ocean and by outgassing in relatively warm upwelling water near the equator.

Some other examples: sea ice hinders the exchanges between atmosphere and oceans; the biosphere influences the carbon dioxide concentration by photosynthesis and respiration, which in turn is influenced by climate change. The biosphere also affects the input of water in the atmosphere through evapotranspiration, and the atmosphere's radiative balance through the amount of sunlight reflected back to the sky (albedo).

These are just a few examples from a virtually inexhaustible list of complex interactions some of which are poorly known or perhaps even unknown. Any change, whether natural or anthropogenic, in the components of the climate system and their interactions, or in the external forcing, may result in climate variations.

Koppen's Climate Classification

Of the several schemes of climatic classification, the one devised by Wladimir Koppen, a German botanist and climatologist, still remains the most widely known descriptive system. It is a quantitative as well as empirical classification of climate. Koppen proposed his first classification in 1900, using the world vegetation map of de Candole, a French plant physiologist. This classification scheme uses certain critical values of temperature of the warmest and the coldest months and precipitation of the wettest and the driest months.

It uses numerical values for delimiting the boundaries of different climatic regions and types. Koppen's climatic regions in many cases coincided with the distinct vegetation regions. The plant classification proposed by A. de Candole in 1874 was accepted by Koppen. The following are the five principal biological groups that are largely controlled by temperature and moisture.

Megatherms

This group includes plants which require uniformly high temperature and abundant supply of moisture. The season is winterless, the average temperature of the coldest month being above 18 °C. There is at least one month of heavy precipitation. Certain areas of this belt are characterized by two rainy seasons. The characteristic vegetation is the tropical rain forest.

Xerophytes

Plants that prefer aridity and need high temperatures, even though for a short season, are classed as Xerophytes. This class of vegetation is found in the semi-arid steppes and hot deserts. These plants are also found in the warmer parts of the middle latitude zone. The type of vegetation varies with the soil types.

Mesotherms

This group of vegetation consists of plants that are adapted to moderate heat and a moderate amount of moisture. Certain types of plants are not adapted to low winter temperatures, while others shun dryness of the warm season. These plants are found in regions lying between latitudes 22 ° and 45 °N and S. The average values of temperatures are usually below 18 °C for the coldest month and 22 °C for the warmest month.

Microtherms

Plants that need lower mean values of annual temperature, summers cool and short, and winters colder are called microtherms. The monthly mean temperature for the warmest month is at least 10 °C and less than 22 °C.

The mean temperature for the coldest month is below 6 °C. Occasional snow in winter and adequate precipitation during the warmer months suit the climatic requirements of these plants. Evergreen deciduous forests and steppes are the natural vegetation regime.

Hekistotherms

This group of vegetation comprises of plants of the snow-bound Arctic region, beyond the polar limits of tree growth. Mosses, lichens, etc. are the natural vegetation. Not satisfied with his first scheme of climatic classification, Koppen revised and modified it several times during his own life time.

Koppen revised his classification first in 1918, when he paid greater attention to the monthly and annual averages of temperature and precipitation, and their seasonal distribution. Thereafter several modifications were incorporated in his classification by the author himself.

Koppen-Geiger world climatic map was published in 1936. However, a further modified version of Koppen's original classification was published in 1953, which is known as Koppen-Geiger-Pohl's classification of world climate.

In classifying climate Koppen placed reliance on his belief that the distribution of natural vegetation was the best expression of the totality of climate. Consequently, many of the climatic boundaries he selected were based on vegetation limits. The Koppen system recognizes five principal categories of climate; each category is designated by a capital letter as follows:

- Humid tropical climates: Winterless climates; it is hot all seasons; all months have a mean temperature above 18 °C.
- Dry climates: In these climates evaporation exceeds precipitation; there is a constant water deficiency.
- Humid mesothermal climates or warm temperate rainy climates: These climates have mild winters; the average temperature of the coldest month is below 18 °C but above – 3 °C; the average temperature of the warmest month over 10 °C. In this group of climate both the seasons, winter and summer, are found.
- Humid micro-thermal climates or cold snow-forest climates: These climates have severe

winters; the average temperature of the coldest month is below -3 °C and that of the warmest month exceeds 10 °C.

- Polar climates: These are summer-less climates; the warmest monthly mean is below 10 °C.

It may be noticed that four of the principal categories of climatic groups (A, C, D, and E) are based on temperature characteristics, while the fifth, the B category, has precipitation as its fundamental criterion.

Thornthwaite's Climate Classification

Charles Warren Thornthwaite, an American climatologist attempted classification of world climates in 1931. The climatic classification was revised and improved further in 1948. The classification incorporates evapotranspiration, temperature and precipitation information and is widely used in studying the animal species diversity and potential impacts of climate change. Thornthwaite's classification is an improved and modified form of Koeppen's classification of climates. It is assumed that precipitation, temperature and plant association are strongly correlated. Natural vegetation is the true representation or the product of the climate. It is an outcome of precipitation and thermal conditions prevailing at a specific place over a long period of time. Growth and development of natural vegetation not only depended upon precipitation alone but also on the rate of evaporation. The classification of climates is based on the concept of potential evapotranspiration (PET). By potential evapotranspiration we mean the total amount of water vaporized and transpired over a certain area if sufficient amount of water is available.

Method of Calculation

Thornthwaite used a set of indices to work out classification of climates. The indices used in the exercise are potential evapotranspiration (PET) index, heat index (I), moisture index (MI) and thermal index (TI).The potential evapotranspiration (PET) is calculated from mean monthly temperature (in oC) in association with day length for a 30 day month (12- hour days). Heat index (I) is worked out by monthly mean heat index adjusted for the number of hours of day light and days in specific months. It is linked to mean monthly air temperature. Moisture index (MI) is worked out by subtracting PET values with precipitation (P) divided by PET and the coefficient multiplied by 100. Thermal index (TI) is worked out by adding minimum temperature values to maximum and the sum divided by two. Equation form of each Index is given below:

PET Index

Potential evapotranspiration (PET) index is worked out using mean monthly air temperature in °C.

$$PE=1.6\frac{(10Ta)^a}{}$$

Where,

PE= Monthly potential evaporation in centimeters

Ta=Mean monthly air temperature in ° C

^{a}= is a cubic function of I, given as

^{a}= (6.75x 10^{-7}) x I^3 – (7.75-10^{-5}) x I^2

+ (1.792 x10^{-3}) x I+ 0.49

Heat Index

Heat index refers to the mean monthly air temperature (°C) for each month from month 1st to month 12th in degree Celsius. It is further adjusted to number of hours of day light and days in a month. The total values of temperature were then divided by 5, a constant value of average. The coefficient so derived is raised by the power value of 1.514. Thus, heat index refers to sum total of monthly temperature values adjusted to hours of day light divided by a constant value. The final index is further raised to the power 1.514to ascertain the heat index.

I= Annual heat index, given by

$I = \sum_{N-1}^{n-12} (Ta/5)^{1.514} . (Ta/5)^{1.514} = i$ =Mean heat index of the nth month.

$E_T = 1.60\ L_a \left(10T \div I_T\right)^a$

E_T =monthly PET (cm)

La=adjustment for the number of hours of day light and days in the month (depends on latitude)

T=mean monthly air temperature (°C)

I_T =total of 12 monthly values of heat index $I = \sum_1^{12} i$

i= $(T/5)^{1.514}$

a= empirical cons tan t= 6.75x10^{-7} $(I_t)^3$ -7.71x10^{-5} $(I_t)^2$ + 1.792 x10^{-2} (I_t) +0.49239

Moisture Index

The proportion of precipitation that ultimately remains available for practical purposes after evaporation is known as moisture index. It is being worked out bysubtracting the value of potential evapotranspiration (PET) from precipitation and the balance is divided by potential evapotranspiration (PET). The coefficient is then multiplied by 100 to get the moisture index.

(MI) = (P-PET÷PET) x100

Where,

P = Precipitation

PET =Potential evapotranspiration

Thermal Index

The index is being worked out by adding together the maximum temperature value of each day for the month and the sum divided by total number of days in the month. Similarly, average minimum temperature is also worked out by adding together diurnal minimum temperature values for a month and the sum divided by the total number of days in the month. Thermal index is calculated by the addition of average maximum temperature and average minimum temperature and the sum divided by two. Thermal index is, thus, the middle most value lying between maximum and minimum values of temperature.

$$(TI) = (Tmax + Tmin) \div 2$$

Where,

T_{max} = Maximum temperature

T_{min} = Minimum temperature

Monthly water surplus (S) or deficit (D) is calculated from a moisture budget assessment including stored soil moisture. A scheme of moisture index (MI), humidity province (expressed in capital alphabets), rainfall (in centimeters), precipitation effectiveness (PE) based on heat (I) and moisture indices (M) and thermal province (expressed in capital alphabets with superscript dash) is presented in Table. Heat and thermal indices are extremely useful in working out precipitation effectiveness, which in its turn, is an important indicator of plant growth. Heat index refers to hours of sun light and air temperature whereas thermal index relates to the average thermal condition, i.e. sum total of maximum and minimum temperature divided by two. The table below summarizes different indices and their limits.

Table: Major Parameters of Climatic Classification.

Moisture index	Humidity province	Rainfall (in Cm)	PE/I m	Thermal province
100 and above	Pre –humid (A)	114	44.9	Tropical or Mega thermal (A')
20-100	Humid (B_1 to B_4)	57-114	22.4-44.9	Meso thermal(B'_1 to B'_4)
0 to 20	Moist sub-humid (C_2)	28.5-57	11.2-22.4	Micro thermal (C_1 to C_2)
-20 to 0	Dry sub-humid (C_1)	14.2-28.5	5.6-11.2	Steppe
-40 to -20	Semi- arid (D)	<14.2	<5.6	Taiga (D')
-60 to – 40	Arid (E)	<14.2	<5.6	Tundra (E')Frost(F')

Thornthwaite's classification is, thus, based on two important indices:

- Precipitation Effectiveness (PE): Effectiveness of precipitation depends on the distribution pattern of rainfall in a calendar year and evaporation of moisture back to the atmosphere through different surfaces and means. Seasonal deficiency and surplus of rainfall affects plant growth significantly.
- Temperature Efficiency (TE): Temperature efficiency relates to variation in temperature and its role in the growth of plants. Equatorial areas where sunlight and temperatures are

usually high with insignificant variation in a calendar year, growth of plants remains high. Contrary to it, arid areas reflect high variation in temperature and sun-light, hence plant growth is retarded significantly. Potential evaporation is, thus, an index of temperature efficiency.

Design of Classification

Thornthwaite's design of climate classification is a combination of three letter alphabets. First alphabet used in the major climatic classification is any one of the English capital letter from A to E which denotes world's five moisture regimes. Further, sub division in a major climate is expressed by small numerical values such as 1, 2, 3 denoting variation within moisture regimes. These are:

A =Wet or per humid (with moisture index as 100),

P/E index= 127 and type of vegetation 'rainforest'.

B = Humid (with moisture index 20 to 100) B_1, B_2, B_3 and B_4,

P/E index= 64-127 and type of vegetation 'forest'.

C= Moist sub humid (with moisture index 0 to 20),

P/E index =32-63 and type of vegetation 'grasslands'.

D = Dry sub humid (with moisture index -30 to 0),

P/E index= 16-31 and type of vegetation 'steppe.

E =Arid (E with moisture index -100 to -67),

P/E index=<16 and type of vegetation 'desert'.

The second letter used in the climatic classification is also an English capital alphabet superscript with dash. It denotes thermal provinces. Thornthwaite has divided the world in to six thermal provinces. They are expressed as:

A'= Mega thermal (Tropical), T/E index= 127.

B'= Meso thermal expressed as B1, B2, B3 and B4.

T/E index= 64-127.

C'= Micro thermal, T/E index = 32-63.

D'= Taiga T/E index = 16-31.

E'= Tundra T/E index = 1-15.

F'= Frost T/E index = 0.

Based on the moisture and thermal indices, 9 major climates were identified.' Third letter in combination of alphabets is denoted by a set of 8 small English alphabets. They are used to express the seasonal pattern of the availability of moisture from acute deficit to abundant surplus situations. The expression of small alphabets is explained below:

Rainfall adequate in all seasons = r.

Rainfall deficient in all the seasons = d.

Rainfall deficient (normal) during winter season = w.

Rainfall deficient (normal) during summer season = s.

Aridity Index for Humid Climates

Moisture deficit acute during winter = w_2.

Moisture deficit acute during summer =s_2.

Humidity Index for Arid Climates

Moisture surplus abundant during winter =s_2.

Moisture surplus abundant during summer = w_2.

Thornthwaite has been concerned about working out seasonal deficiency of moisture in humid climates which appears to be the cause of dryness and apparently halting the growth of plants for specific period. Similarly, places in arid climates do represent moisture surplus seasonally which serves as storage of soil moisture and a basis for growth of plants as grasslands and xerophytes.

Thus, precipitation effectiveness (PE) and temperature efficiency (TE) are closely related and are the expressions of vegetation type. Both the indices record high values where variations in the distribution of pattern of rainfall and temperature reflect little variation. Contrary to this, both the indices record low values if the distribution pattern of rainfall and temperature reflect high variation. However, both the indices are aimed at a common point of reference i. e. vegetation which is true representation of a climate type.

The third alphabet classification is largely based on the seasonal characteristics of precipitation. The following four small alphabets were used to denote the seasonal characteristics of rainfall. Based on four seasonal characteristics of rainfall, 20 climatic divisions have been recognized. They are presented in Table.

Table: Classification of Climates Based on Moisture Regimes and Seasonal Characteristics of Moisture.

Moisture regimes / Seasonal pattern of moisture	Wet or Per Humid	Humid	Moist Sub Humid	Dry Sub Humid	Arid
Adequate rainfall in all the seasons	Ar	Br	Cr	Dr	Er
Rainfall deficient in summer	As	Bs	Cs	Ds	Es
Rainfall deficient in winter	Aw	Bw	Cw	Dw	Ew
Rainfall deficient in all the seasons	Ad	Bd	Cd	Dd	Ed

Besides surplus and deficient rainfall at seasonal level, Thornthwaite has also considered the level of seasonal moisture deficiency for humid climates and seasonal abundance of moisture for arid climates as crucial elements in plant growth and the characterization of climates. As such, an attempt was made to work out season specific moisture deficiency from normal to acute for humid climates and moisture surplus from normal to abundant for arid climates. To make it distinct from normal situation, seasonal characteristics were further assigned with numerical values.

$$M = Ih - 0.6\ Ia$$

Or,

$$M = \frac{100\ \alpha\ -\ 60\ \beta}{\gamma}$$

Where,

Ih is the index of humidity,

Ia is the index of aridity,

Alpha (α) is the water surplus,

Beta (β) is the water deficit,

Gamma (γ) is the water need or potential evapotranspiration.

The above concept relates to the status of moisture at a specific place. If alpha (α) exceeds the potential evapotranspiration (PET) or water need (γ), the situation is termed as humid. Similarly, if beta exceeds the potential evapotranspiration (PET) or water need (γ), the situation is termed as arid. The seasonal characteristics are explained below:

Aridity Index for Humid Climates A, B, and C_2

Proportion of moisture deficit is denoted by:

No moisture deficit 0-10 = r.

Moisture deficit (normal) during summer season 10-20 = s.

Moisture deficit (normal) during winter season 10-20 = w.

Moisture deficit acute >20% in winter = w_2.

Moisture deficit acute >20% in summer =s_2.

Humidity Index for Arid Climates C_1, D and E

Proportion of moisture surplus is denoted by:

Water surplus negligible 0-16.7 =d.

Winter moisture surplus (normal) 16.7-33.3=s.

Summer moisture surplus (normal) 16.7-33.3=w.

Moisture surplus abundant >33.3% in summer =w_2.

Moisture surplus abundant >33.3% in winter =s_2.

The seasonal variation of effective moisture and summer concentration of temperature efficiency has been further refined for the climatic classification.

Table: Thermal Efficiency and its Summer Concentration.

Thermal efficiency	Climate type	Index value	Type of summer concentration	Percentage of concentration
A^1	Mega thermal	>114	A	< 48.0
B^1_4	Meso thermal	99.7 – 114	b_4	48 - 51.9
B^1_3	Meso thermal	85.5 - 99.7	b_3	51.9 - 56.3
B^1_2	Meso thermal	71.2 – 85.5	b_2	56.3 - 61.6
B^1_1	Meso thermal	57.2 – 71.2	b_1	61.6 - 68.0
C^1_2	Micro thermal	42.7 – 57.2	c_2	68.0 - 76.3
C^1_1	Micro thermal	28.5 - 42.7	c_1	76.3 - 88.0
D^1_1	Tundra	14.2 - 28.5	D	>88.0
E^1	Frost	<14.2	E	>88.0

Water Balance and its Explanation: There is a close relationship between rainfall, temperature and potential evapotranspiration, though they are not linked to a common cause and are evidently the result of different causes. The potential evapotranspiration rates increase with an increase in temperature and decrease in relative humidity which varies with precipitation. In certain situation, rates increase steadily from January to May, with increasing temperature and aridity, followed by a rapid decline from June onwards with the onset of summer monsoon. The minimum is ultimately reached in the month of August which also records the heaviest rainfall and resultant humidity. The effect of relative humidity on the rates may be seen from the examples of the months April and July which have mean maximum temperatures of 36° C and 34° C. respectively and the corresponding potential evapotranspiration rates of 30 and 15 centimeters. A gradual rise in rates is noted from September with a slight increase in air temperature following a decline in rainfall and cloudiness. However, the rise in rates does not continue beyond October because of approaching cold season. The actual evapotranspiration, as distinct from potential evapotranspiration, depends upon the amount of water actually available for the growth of plants. As such, it is a function of precipitation provided there is no soil moisture recharge by capillary action from the water table and that artificial irrigation is not prevalent. Further the actual evapotranspiration in any month will be less than or equal to the corresponding potential evapotranspiration according to an existence or otherwise of a water deficit in that month. As such, actual equals the potential only in July, August and September when water deficit is non-existent.

Soil-Water-Balance (SWB)

Thornthwaite and Mather developed a method for estimating the recharge, storage and abstraction in the process of water percolation to subsoil layers. Thornthwaite as a climatologist, developed the concept of the soil-moisture balance, and he preferred to use it as the foundation of his climate system. He was convinced that soil moisture balance represents availability of moisture for the plants and an assessment of the availability surplus moisture to supply stream flow and ground water. This concept implies that precipitation alone does not indicate the amount of water actually available to plants. The moisture requirement of plants becomes higher as temperature increases. A schematic figure is presented below. It illustrates that rain and irrigation is the two sources of the soil moisture. Evaporation and transpiration are the two ways of moisture transfer to atmosphere. Runoff and percolation are the processes of water flow. Stored soil water has deep percolation and upward movement of soil moisture.

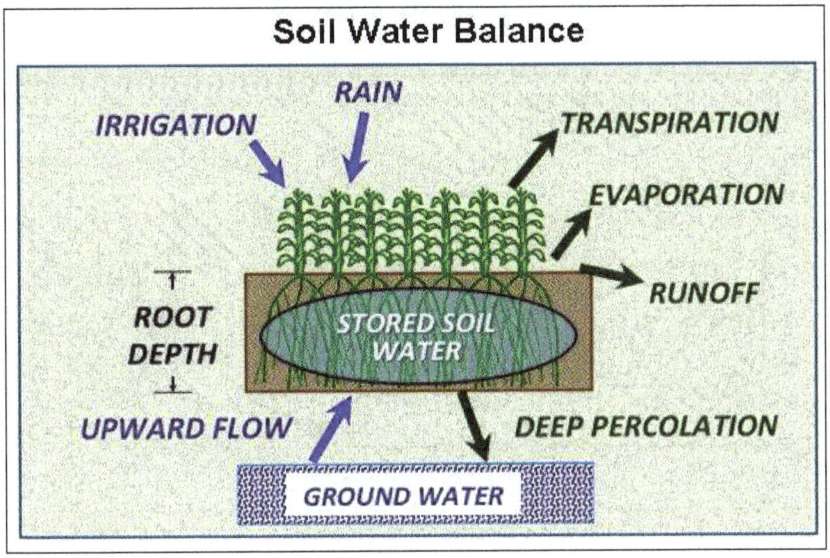

Figure: Soil-Water –Balance.

The method developed by Thornthwaite is given below:

Recharge= (Precip+ snow melt +inflow) – sources (Interception+ Outflow +ET)-▲ soil moisture sinks.

$$R = \frac{(P - I_a)^2}{(P + (S_{max} - I_a)}$$

Where,

R is run off,

P is daily precipitation,

S_{max} is maximum soil moisture holding capacity,

I_a is initial abstraction, the amount of precipitation that must fall before any runoff is generated.

The initial abstraction (I_a) term is related to a maximum storage term (S_{max}) as follows:

$I_a = 0.2\ _{Smax}$

The maximum storage term is defined by the curve number for the land- cover type under consideration:

$$S_{max} = \underline{(1000)}\ \text{-10/CN}$$

Thus, it has been established that availability of water is crucial to the survival and growth of plants and animals including humans. Currently, the world is faced with water crisis and declining bio diversity. Thornthwaite could perceive the prospective crisis of water nearly seventy years before and for that he developed the method and a diagram to represent the situation so that possible solutions could be searched out. Figure represents a water balance diagram.

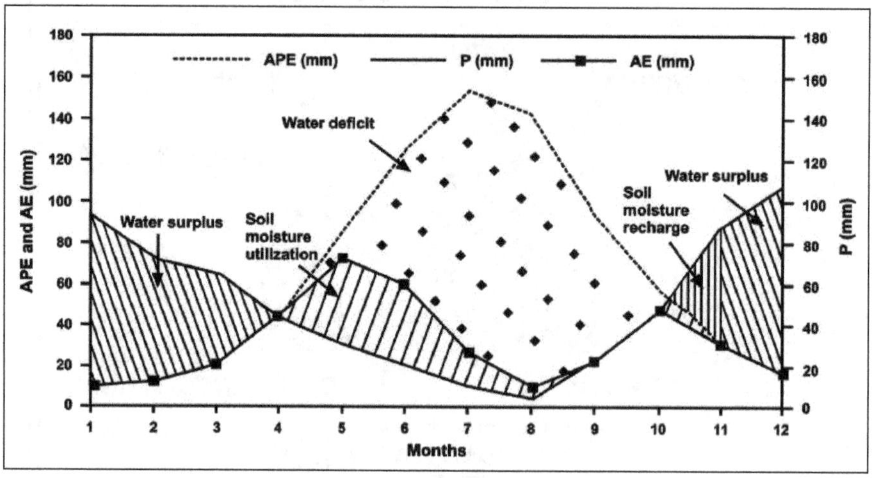

Figure: Plotting of Water Balance.

Three parameters PE, TE and seasonal characteristics of surplus and deficit moisture conditions formed the basis for dividing world into 32 types of climates. Moisture provinces A and E have three subdivisions in each. Similarly, B and D moisture provinces have seven and six sub divisions each while C moisture province has a maximum of 10 sub divisions. Thermal provinces D', E', and F' have been represented by single climate type without having further subdivisions in them. Major climate types have been presented in following Table.

Table: Major climate types (based on moisture regime, thermal province and seasonal characteristics).

Per humid	Humid	Moist sub humid	Dry Subhumid	Semiarid	Arid	Arid	
Mega thermal	Meso thermal	Micro thermal	Steppe	Taiga	Tundra	FrostFrost	
AA'r	BA'r	CA'r	DA'w	EA'd	D'	E'	F'
AB'r	BA'w	CA'w	DA'd	EB'd			
AC'r	BB'r	CA'd	DB'w	EC'd			
	BB'w	CB'r	DB's				
	BB's	CB'w	DB'd				
	BC'r	CB's	DC'd				
	BC's	CB'd					
		CC'r					

		CC's					
		CC'd					
3	7	10	6	3	1	1	1

Note: The first row of the table relates to titles: Moisture regimes (bold letters) and thermal provinces (italicized) for each column. A workable scheme for identifying the type of climate based on potential evapotranspiration, average annual thermal efficiency, seasonal variation of effective moisture, and summer concentration of thermal efficiency has been presented by Thornthwaite. A map accessed from internet is also presented below. It is easy to recognize polar climates at the top closely followed by tundra type climate towards south. Close to equator, one can notice A, B, and C type of climates given successively northward. Desert climates (D type) are recognizable along the margins of tropical and subtropical belts. The map showing climatic divisions of the world, as per Thornthwaite's classification, is presented in figure.

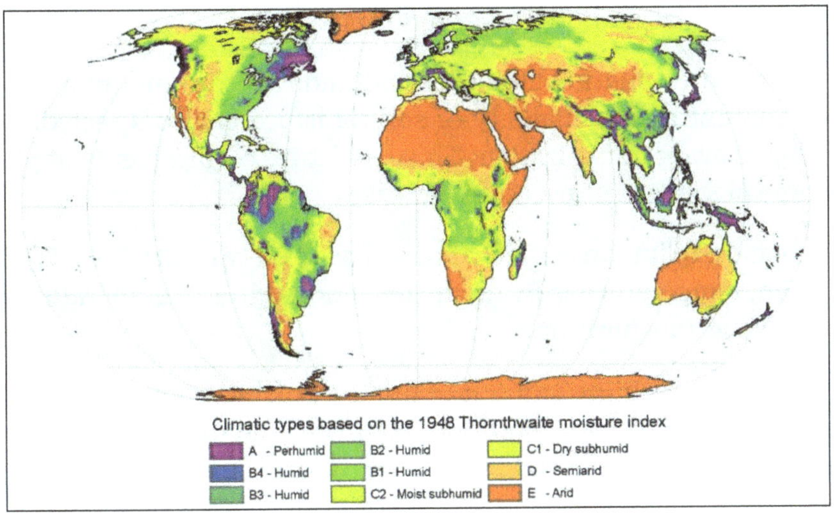

Spatial distribution of world climates: The pattern of spatial distribution of climates reveals a pattern of zonation from equator to poles. While equatorial areas with high P/E and T/E index values express luxuriant vegetation in the form of rain forests, Tundra and Polar areas with low P/E and T/E indices express scanty and limited vegetation. Similarly, eastern coasts in subtropical areas have high P/E values and moderate T/E values and thus have forests which are mostly deciduous in nature. While humid climates (A, B, and C) are characterized by normal seasonal deficiency of moisture, arid climates (D & E) do have normal seasonal surplus of moisture in at least one season. Thus dry climates usually occupy a large place in interior parts of the continents, the western margins of the continents in tropics and eastern margins of the subtropics and temperate zones. London (U.K.) is an example of humid oceanic climate, while Athens (Greece) is an example of mild winters in Mediterranean climate. Lima (Peru) represent dry continental climate while Denver in (U.S.A.) humid continental climate. La Paz (Bolivia) at an altitude of 3664 meters above sea level has very low P/E and very low T/E index. It represents highland climate.

Criticism of Thornthwaite's 1948 Classification

This scheme of climatic classification has been successfully applied to many regions, but because of certain limitations inherent in it no world map has yet been prepared. This classification system has proved most satisfactory in case of North America where vegetation boundaries nearly coincide with particular PE values. But it is not satisfactory for the tropical and semi-arid areas. However, Thornthwaite does not determine his climatic boundaries on the basis of vegetation boundaries. In this respect, he differs from Koppen in case of his second classification.

This classification is quantitative as well as empirical, but it does not take into account various causative factors of climate. It also employs symbols for designating the climatic types. The classification involves a lot of calculations, hence it is more difficult to use in determining the climatic type of a particular place or locality. That is why Thornthwaite's system could not enjoy the same degree of popularity as that of Koppen.

Lastly, Thornthwaite as a climatologist developed the concept of the soil-moisture balance, and he preferred to use it as the foundation of his climate system. He was convinced that the soil- moisture balance represents availability of moisture for plants, and an assessment of the availability of surplus moisture to supply stream flow and ground water.

This concept implies that precipitation alone does not indicate the amount of water actually available to plants. The amount and availability of soil moisture is also affected, besides other factors, by the losses due to evapotranspiration. Thus, moisture requirements of plants, according to Thornthwaite, become higher as temperature increases.

Even though this scheme of climatic classification is based on two variables: (1) precipitation, and (2) evapotranspiration, temperature is by no means ignored in it. Temperature is also accounted for while calculating evapotranspiration.

If a comparison is made between the amount of water available from precipitation and the water need, it is easy to assess the moisture conditions to determine the seasonal distribution of moisture deficits or surpluses and whether a climate is wet or dry.

The most significant contribution of Thornthwaite has been that the concept of evapotranspiration has been applied in practical studies of the water balance as regards the problems of water use. However, in spite of numerous plus points of Thornthwaite's classification of the climate, his system is seldom used in an introductory course because it involves a lot of very complicated calculations.

Alisov's Climate Classification

To analyze the patterns of formation of climates within the global system and to solve practical problems, it is necessary to know the distribution of climatic values above and over the globe or region, as well as the climate complex as a whole.

There are different approaches to the classification of climates. If this is done for the purpose of analyzing the origin of the climate itself or for linking with a complex of natural conditions (landscape-geographical zones), then such a division of climates is called climatic classification., and if for applied purposes (servicing agriculture, construction, transport) - climatic zoning.

Climate classifications and regionalization are numerous and determined by different tasks. There are classifications that link the distribution of vegetation, soil, river network, relief in general to climate, or study the patterns of formation of the global climatic system from local climates.

Modern classifications and regionalization's are not limited to the division of climates, they also reveal their system, thereby paying attention to their similarities. The genetic classification of climates is based on the division of the earth's surface into climatic zones and regions in accordance with the conditions of general atmospheric circulation, expressed in the prevalence of air masses of a certain geographical type, during the year or in one of the two main seasons. In addition to the seasonality of circulation conditions, there are two types in each zone: lowland climate and highland climate. This provides a basis for linking circulation boundaries with landscape boundaries.

In real life, the situation is more complicated. Circulation processes determine humidification at low latitudes, and thermal conditions differ little, and therefore the climate boundaries according to Alisov coincide well with landscape zones. In temperate latitudes, moisture is also determined by atmospheric circulation. When determining the climatic boundaries, the distance of various parts of the continent from the ocean was taken into account. The situation with thermal boundaries in the warm half of the year, which in the extra tropical zone strongly depends on radiation conditions, is much more complicated.

The formal boundaries of the propagation of air masses under conditions of their continuous thermal transformation do not always correspond to the landscape boundaries. Latitudinal zones and types of climates according to Alisov, their features are given in the table. Latitudinal climatic zones represent four zones, where someone air mass (EV, TV, HC, AB) prevails, and three zones, where air masses of lower latitudes prevail in summer and higher latitudes in winter.

B.P. Alisov identifies seven main climatic (circulation) zones: equatorial, two tropical, two temperate, arctic and Antarctic. Each zone is characterized by a constant predominance of air masses of a geographic type of the same name as the zone. Then, intermediate zones are distinguished: two zones of equatorial monsoons with a winter prevalence of tropical and summer equatorial air, two subtropical zones with a winter prevalence of polar and summer tropical air, a subarctic zone with a winter prevalence of Arctic air and summer air of temperate latitudes. Scheme of latitudinal zones (zones) and types of climate (according to B.P. Alisov).

Alisov took as a basis important circulation signs that integrally characterize the temperature and humidity state of the climate of a particular region. He used the predominance of certain air masses in different seasons as a defining indicator, and the most frequent positions of the main atmospheric fronts (climatic fronts) were taken as the climate boundaries.

There are no quantitative characteristics for the position of these fronts, so they are roughly based on synoptic experience. Let us give a brief overview of the Alisov climate system, presented in table. In each hemisphere, four climatic zones are distinguished according to the predominance of one of the main air masses in them during the year: equatorial, tropical, temperate, and arctic (Antarctic).

Between the main zones, there are three transition zones, where, due to seasonal migration, air masses of lower latitudes prevail in summer and higher latitudes in winter. It is a subequatorial belt, in which equatorial air prevails in summer and tropical air in winter. In the subtropical zone, tropical air dominates in summer, moderate in winter. In the subarctic (sub Antarctic) belt, the air is temperate in summer and arctic (Antarctic) in winter. In each of the latitudinal belts, continental and oceanic types of climate are distinguished. In addition, due to the fundamental differences in circulation processes at different boundaries of the continents, the types of climate of the eastern and western coasts of the continents are distinguished, including both part of the land and part of the adjacent sea area. A mountain type of climate is distinguished in each belt.

Table: The system of belts and zones according to B.P. Alisova.

1.Equatorial	2.Subequatorial (tropical monsoon climate)	3.Tropical	4.Subtropical	5.Moderate	6.Subpolar climate (subarctic and sub Antarctic)	7. Climate of the Arctic (Antarctic)
Continental	Continent. monsoons	Continental (3a)	Continental (4a)	Continental (5a)	Continental	Arctic climate (7a)
Oceanic	Oceanic. monsoons	Oceanic (3b)	Oceanic (4d)	Oceanic (5c)	Oceanic	Antarctica climate (7b)
	Monsoons of the west shores	The climate of the eastern periphery is oceanic Az	Western coastal climate (Mediterranean) (4b)	The maritime climate of the western shores (5 B)		
	Monsoons of the eastern shores	The climate of the western periphery is oceanic Az	(4c)	Climate of the eastern shores (monsoon) (5g)		

Belts 1, 3, 6 do not appear on all meridians, the first two are due to large shifts in circulation over the seasons, the latter, on the contrary, due to the absence in some areas of a significant seasonal shift in the boundaries of air masses. According to B. P. Alisov's classification, continental and oceanic types of climate are distinguished in the equatorial climate. They do not differ too much among themselves in terms of temperature and humidity. Over the continent at equatorial latitudes, the air is also very humid due to the high moisture content of the underlying surface and the large evaporation of tropical vegetation. Therefore, no distinction is usually made between oceanic and continental equatorial climates. In latitudes close to the equator (up to 5–10 ° in each hemisphere), where the influx of solar radiation varies little throughout the year, a very uniform temperature regime is observed. Both at sea and on land, the average long-term temperatures of all months of the year are from +24 to +28 °C. The annual temperature range can be no more than 1 °C and usually does not exceed 5 °C. The daily temperature amplitudes are of the order of 10–15 °C. Evaporation is great, therefore the absolute humidity is also great. It can exceed 30 g / m 3. The relative humidity is also great. Even in the driest months of the year, it is above 70%.

Precipitation in the equatorial type of climate is plentiful, heavy and often accompanied by thunderstorms. Most of them fall in the intertropical convergence zone: over the sea, they are not as intense and not as frequent as over land. In general, 1000-3000 mm of precipitation falls per year.

But in a number of places, for example, in the mountainous regions of Indonesia, in Central Africa, more than 6000 mm falls. In most regions, the distribution of precipitation throughout the year is more or less uniform; on land these are areas of tropical rainforests in South America, Africa, and Indonesia. The landscape is a humid equatorial forest.

In some parts of the tropical oceans, in particular in the Indian and western Pacific oceans, as well as over South Asia and in the tropics of Africa and South America, the tropical monsoon regime dominates. Along with a more or less sharp seasonal change in the prevailing air currents, there is also a change in tropical air to equatorial air from winter to summer. The air temperature in the tropical monsoon zone over the ocean is as high and has the same low annual amplitude as in the equatorial climate. Over land, the annual temperature range is greater and increases with latitude.

The most prominent feature of this type of climate on the continents is the dramatic difference between dry and rainy periods. During the winter, this type of climate is dominated by continental tropical air masses with high temperatures and no precipitation. In summer, with the arrival of the summer southwestern monsoon, which carries humid equatorial air masses, a rainy period begins, the temperature decreases slightly. The amount of precipitation can vary greatly depending on the distance of the point from the ocean, on the latitude, the duration of the rainy period, the orography conditions, the vertical thickness of the equatorial air mass, and other factors.

Thus, in the type of continental tropical monsoons, the year is sharply divided into dry (winter) and rainy (summer) periods. The annual temperature variation reveals a main temperature maximum in spring, a main minimum in winter, a secondary maximum in autumn, and a secondary minimum in summer during the summer monsoon. Due to the long dry period, the typical landscape in this type of climate is the savanna, which transitions to the pole-facing periphery in the steppe and semi-desert.

In this type, as well as in the continental, there is a seasonal change in air masses. In the winter months, tropical marine air prevails here, the properties of which, naturally, differ greatly from the continental air, and above all in temperature and humidity. At the same time, tropical marine air differs little from the equatorial air that comes with the summer monsoon. The type of oceanic monsoons is characterized by small annual (1–2 °C) and daily (not exceeding 2–3 °C) air temperature amplitudes. Average monthly temperatures are 24-28 °C. The most characteristic feature of the climate is the absence of a dry season in the strict sense of the word and the long duration of the summer rainy season. The winter monsoon is northeastern, but since it carries humid tropical marine air, it rains during the winter monsoon, but their amount is much less than during the summer southeastern monsoon, which brings humid equatorial air mass.

This includes the Indian and West African monsoons. The summer monsoon is generally associated with the southwest air currents, the winter monsoon with the northeast. The annual course of precipitation is especially pronounced. In general, we can say that during the dominance of the summer monsoon 75% of the annual precipitation falls. The amount and distribution of precipitation is greatly influenced by the orography of the western shores. So, on high banks and on the slopes of mountains facing the summer monsoon, precipitation increases sharply. In the annual temperature variation, the maximum occurs in spring.

The absolute and relative air humidity (maximum in summer) and cloudiness (sharp maximum in summer and sharp minimum in winter) also sharply change with the seasons. For example, in Kolkata, cloud cover is 84% \u200b\u200bin July and 8% in January.

Due to dry winters, the tropical monsoon climate of the western shores is particularly characteristic of the savannah landscape. One area of \u200b\u200bthe tropical monsoon zone has the highest average annual temperatures on the globe. On the southwestern shores of the Red Sea, Eritrea is sometimes hit by the summer monsoon from the Southern Hemisphere. Its temperature rises further as a result of the foehn process, which is why very high air temperatures are observed on the coast of Eritrea. In Massauja (15.6 °N, 39.5 °E), the average temperature in January and February is + 25 ÷ + 26 °C, in July +35 °C, and the average annual temperature is + 30 °C.

The most characteristic distinguishing feature of this type from the previous one is the long duration of the rainy period, often with a shift in the maximum precipitation to late summer and early autumn and a relatively humid dry period. The air currents of the summer monsoon are southwestern, southern and even southeastern, while the winter monsoon is northeasterly. The northeastern currents of the winter monsoon, before entering the mainland, have time to travel a fairly long way over the water surface of the East China and South China Seas and the western Pacific Ocean, so the winter monsoon is quite humid here.

So, in Ho Chi Minh City (Vietnam, 10 ° 49 "N, 106 ° 40" E), the summer monsoon begins in May (196 mm of liquid precipitation) and ends in November (122 mm), with the main maximum rainfall in September (292 mm), and the secondary maximum in June (285 mm). Of the 1806 mm of rain falling per year, almost 93% falls on the rainy season and 7% on dry. The maximum temperature falls in April (29 °C), during the rainy period the temperature drops from 28 °C in May to 26 °C in October and then does not change during the dry season. The landscape is a humid equatorial forest.

Tropical climates are formed in areas dominated by tropical air masses. It is known that tropical air over the oceans forms in subtropical anticyclones. Above the continents, the region of its formation is the deserts and savannas of the tropical belt. The tropical belt is characterized by high values \u200b\u200bof the radiation balance: over the oceans they amount to about 5000 MJ / m2 year, and over the continents - 2500 MJ / m2 year.

Over the continents in this belt, due to extreme dryness, the heat consumption for evaporation is low. Therefore, despite the large values \u200b\u200bof effective radiation and large albedo, the surface of the continents warms up strongly in summer and the continental tropical air acquires high temperatures.

This climate is observed in North and South Africa, in Arabia, in most of Australia, in Mexico, in the middle of South America, that is, in those areas on both sides of the equator where there is no monsoon change, where tropical air prevails all year round. These areas make up the zones of tropical deserts, which include the Sahara, the Arabian Desert, the deserts of Australia, etc.

Cloudiness and precipitation are very low here; the radiation balance of the earth's surface is less than in the equatorial belt due to the dry air and high albedo of the earth's surface. However, the air temperature is very high, since the heat consumption for evaporation is low. Summers are extremely hot, the average temperature of the warmest month is not lower than + 26 °C, and in some places almost up to + 40 °C. It is in the zone of tropical deserts that the highest temperature maxima on the globe are observed - about +57 ÷ + 58 °C. Winter is also warm, with the temperature of the coldest month between + 10 ... + 22 °C.

Daily temperature fluctuations are very large, sometimes reaching 40 °C. Annual precipitation in most cases is less than 250 mm, and in some places less than 100 mm. This type of climate is determined by the properties of tropical marine air, which forms in those latitudes of subtropical anticyclones, to which the intertropical convergence zone does not reach in its seasonal shift.

Along the periphery of the subtropical anticyclones facing the equator, a stable trade wind regime with characteristic weather and climate conditions prevails here all year round. The oceanic tropical climate is characterized by moderately high temperatures that increase towards the equator. Average temperatures in summer months are from + 20 to + 27 °C. In winter, in the high-latitude parts of the trade wind zone, temperatures drop to + 10 ... + 15 °C. The presence of an anticyclonic trade wind inversion at a low altitude does not allow convection to develop in height. Cumulus and stratocumulus clouds occur in large numbers: average cloud cover is about 50% or more. But there is no large precipitation in the trade wind zone, with the exception of those islands where orographic conditions favor them. In the open ocean, the weather in the trade wind zone is predominantly dry, since the clouds do not reach the level of glaciation. Light precipitation associated with the coagulation of droplets in clouds does not significantly change this situation. A characteristic feature of this type of climate is the development of tropical cyclones over the oceans, excluding the southern part of the Atlantic and the southeastern part of the Pacific oceans. In addition, they are associated with significant precipitation in this generally extremely dry climate.

It is also called the climate of the western coasts of the continents. On the western coasts of the continents, the winds of the trade winds of the northern quarter in the Northern Hemisphere and the southern quarter in the Southern Hemisphere prevail throughout the year. In the trade wind zone, temperatures are relatively low, since air quickly enters here from high latitudes along the eastern periphery of subtropical anticyclones and, in addition, flows over the cold waters of cold ocean currents. The annual temperature range is small, as well as over the oceans. There is very little precipitation (less than 100 mm per year) due to low water temperatures and low-lying trade wind inversion, but the humidity is high (80-90%) and fogs often appear. This is the climate of coastal deserts such as the Western Sahara coast, southern California, the Namib (South Africa) and the Atacama (South America).

(The climate of the eastern coasts of the continents). This type of climate is very different both from the climate of the eastern periphery of oceanic anticyclones and from the continental tropical climate. Throughout the year, on the eastern coasts, the transport of marine tropical air by trade wind winds prevails. The trade winds on the western periphery of subtropical anticyclones, passing a long way over the oceans, are enriched with moisture and acquire components directed towards the poles.

Due to the prevalence of tropical marine air in all seasons of the year, a warm, humid climate with a large amount of precipitation forms on the eastern coasts. The climate of the eastern coasts of tropical latitudes somewhat resembles the equatorial climate, but differs from the latter in large annual air temperature amplitudes.

The coastal landscape is rainforest. In subtropical latitudes, that is, at latitudes of the order of 25-40 °, climatic conditions are determined by a sharp seasonal change in circulation conditions and, therefore, by a sharp change in the prevalence of air masses. In summer, high pressure zones and polar fronts shift to higher latitudes. In this case, the subtropics are captured by tropical air from lower latitudes. Due to the displacement of subtropical anticyclones to high latitudes, the pressure in the subtropics over the oceans is increased in summer. It is lowered over land. In winter, the polar fronts shift to low latitudes, and therefore the subtropics are captured by temperate air. At the same time, cyclonic activity spreads over the ocean to the subtropics. Over the continents, the pressure in the subtropics in winter increases on average, since the winter continental anticyclones extend to the subtropical latitudes of the cold continents. Hence the seasonal differences in the weather regime, and therefore in the climate of the subtropics.

In summer, the subtropics inside the continents are under the influence of eroded areas of low pressure, the so-called thermal depressions. They form masses of continental tropical air with high temperatures, low moisture content and low relative humidity. The weather here is slightly cloudy in summer, dry and hot. Average temperatures in summer months are close to + 30 °C or exceed this value. In winter, these areas are subject to cyclonic activity, as cyclones often form here or polar fronts pass. The weather is unstable, with sharp changes in temperature and precipitation. Annual precipitation is not more than 500 mm, and sometimes much less. This is a zone of steppes, semi-deserts and deserts.

The large (southern) part of the deserts of the Turan lowland, located in the territory of Central Asia, also belongs to the zone of the continental subtropical climate. Moderate air prevails here in winter, so winter temperatures are much lower than in tropical deserts; in summer, masses of continental tropical air are formed with very high temperatures, low relative humidity, and very little cloudiness and high dustiness. There is no such type of climate in Africa and Australia.

A special kind of continental subtropical climate is the climate of high subtropical highlands. It is observed in the highlands of Asia - Tibet and the Pamirs, at altitudes of 3500-4000 m. The climate is sharply continental, with relatively cool summers and cold winters. There is generally little precipitation; it is a high desert climate. However, the characteristic features of a continental subtropical climate - the predominance of moderate air masses in winter, and tropical air masses in summer - persist here.

In the subtropical latitudes of the oceans, the anticyclone regime with low cloudiness and dry weather with weak winds prevails in summer, and in winter, cyclonic activity with rains and strong winds, often with storms. The annual temperature amplitudes are less than in the continental type, on average about 10 °C. Summer is relatively cool in the eastern parts of the oceans, since air currents from higher latitudes (along the eastern periphery of subtropical anticyclones) often penetrate here, and cold ocean currents pass here. Summers are warmer in the western oceans. In winter, on the contrary, cold air masses from the continents (Asia, North America) enter the western parts of the oceans, and the temperatures are lower here than in the east. In the central parts of the oceans of the Northern Hemisphere, the average temperatures of summer months in the subtropics are + 15 ... + 25 °C, winter months are +5 ... + 15 °C. In the Southern Hemisphere, winter temperatures are higher, summer temperatures are lower, and the annual amplitudes are even smaller.

In summer, the western shores of the continents in the subtropics are located on the eastern periphery of subtropical anticyclones or in their spurs. The weather is clear and dry. In winter, a polar front with cyclonic activity and precipitation often passes or appears here. Thus, the summers in the Mediterranean climate are quite hot and dry, the winters are rainy and mild. Snow may fall annually, but no snow cover is formed. In general, precipitation is not very large. Low precipitation in summer gives the climate a somewhat arid character. This type of climate is typical primarily for the Mediterranean region. Western Transcaucasia belongs to the subtropical type of climate. The Mediterranean type of climate is observed on the coasts of California, Oregon and Washington in North America, on average Chile, in southern Australia, in the extreme south of Africa (Cape Peninsula).

The Mediterranean climate is characterized by distinctive vegetation with many drought-resistant species; these are forests and shrubs with a large admixture of evergreen species. In winter, these areas are influenced by cold northwestern air currents from the mainland; in summer, air from the ocean comes to the southeastern streams. Annual precipitation is opposite to the Mediterranean type. In winter, the weather is clear and dry; in summer, on the contrary, precipitation is abundant, falling in cyclones over the mainland, partly convective, partly frontal. The orographic intensification of precipitation on the windward slopes also plays an important role. Precipitation is generally abundant, and therefore the type of climate is characterized by richly developed broad-leaved forests. Snow falls, but the snow cover does not hold. For example, in Beijing (39.9 °N, 116.5 °E) the average temperature in July is +26 °C, and in January −5 °C; the annual precipitation is 612 mm, of which 235 mm falls in July and 2 mm in December.

In the east of North America, in Washington (38.9 °N, 77.0 ° W), the average temperature in July is + 25 °C, in January + +1 °C; precipitation falls 1043 mm, of which 110 mm in July and 65 mm in November. This type of climate is observed in South America. Buenos Aires (34.6 ° S, 58.5 ° W) has an average temperature of +23 °C in January and +10 °C in July; the annual precipitation is 1008 mm, of which 116 mm in March and 60 mm in July.

In temperate latitudes, there are large seasonal differences in radiation conditions. In summer, the radiation balance of the underlying surface is high and in areas with little cloudiness approaches the conditions of tropical latitudes; in winter on the continents it is negative. Moderate latitudes are characterized by the most intense cyclonic activity at the polar and arctic fronts; therefore the weather regime is very variable here. In the Northern Hemisphere, large differences are observed in the circulation conditions on the continents and oceans, which determine sharply expressed types of marine and continental climates. In the predominantly oceanic Southern Hemisphere, the continental type of climate is practically absent in temperate latitudes.

Very significant differences in climate exist on the western and eastern margins of the temperate continents. The climate of the western shores, subject to the predominant influence of sea air masses, is transitional from maritime to continental; it is often referred to simply as sea. The eastern shores have a monsoon type of climate, especially in Asia.

This type of climate is found on the continents of Eurasia and North America. It is characterized by warm summers and cold winters with a stable snow cover. The annual temperature range is large and grows with distance inland. Humidification conditions vary both from south to north and from west to east.

In the southern part of the temperate latitudes of Eurasia, the high pressure regime prevails in winter. On climatological maps, it is in these latitudes that the center of the Asian winter anticyclone is located. Therefore, winter precipitation is small here and decreases with distance inland. For this reason, the snow cover is low, and in Transbaikalia, near the center of the anticyclone, it reaches negligible values, despite a very harsh winter. In summer, in the southern part of the temperate latitudes of Eurasia, anticyclones of a subtropical type are also frequent, contributing to hot and dry weather. Summer precipitation is higher, but it is also small due to high evaporation at high summer temperatures, so moisture in the southern part of the temperate latitudes is insufficient. In general, 200-450 mm of precipitation falls here per year. As a result, from Moldova, through Ukraine, the southern part of the European territory of Russia and further beyond the Urals, to Mongolia, inclusive, there are steppes, in which dry conditions are often created in summer.

In the higher latitudes of the temperate zone of Eurasia, summers are less hot, but still very warm, winters are more severe (for each given longitude), and annual precipitation is large (300-600 mm). Continentality also grows from west to east (mainly due to lower winter temperatures): annual temperature amplitudes increase and precipitation decreases. The snow cover is higher here and lasts longer. This is an area of \u200b\u200bmixed or deciduous forests. In Moscow (55.8 °N, 37.6 °E) the average temperature in July is +18 °C, in January −10 °C; annual precipitation is 600 mm. In Novosibirsk (55.0 °N, 82.9 °E) in July +19 °C, in January −19 °C; precipitation per year 425 mm. The maximum precipitation is everywhere in the summer.

Further north is the taiga zone, which stretches from Scandinavia to the Pacific Ocean, with the same patterns in climate change from west to east, but with a greater severity of winter. Summers in the taiga zone are as hot as in lower latitudes, but winters are even colder. It is due to the severity of winter that the climate in the eastern part of the taiga zone reaches the greatest continentality. Precipitation falls as much as in the deciduous forest zone; moisture is generally sufficient, in Western Siberia even leading to waterlogging. In Yakutsk (62.0 °N, 129.6 °E) in July +19 °C, in January −44 °C; annual precipitation 190 mm. It is in Yakutia that the continental climate reaches its greatest value. On the mainland of North America, the same types of continental climates of temperate latitudes are found.

The Atlantic and Pacific Oceans occupy large areas in the temperate latitudes of both hemispheres, and the Indian Ocean - in the temperate latitudes of the Southern Hemisphere. The prevailing westerly transport is better expressed over the oceans than over the continents, especially in the Southern Hemisphere. Wind speeds are greater than over continents. In the 40s and 50s latitudes of the Southern Hemisphere, between the zone of subtropical anticyclones and subantarctic latitudes, where the centers of deep cyclones most often pass, the average wind speed is 10-15 m / s. The temperature distribution over the oceans is more zonal than over the continents at the same latitudes, and the differences between winter and summer are less pronounced. Due to the cold summer, the tundra landscape is located on oceanic islands in such low latitudes, in which there is no tundra on the continents. So, the Aleutian and Commander Islands, the Falkland Islands, etc. are covered with tundra.

In the Northern Hemisphere, the western parts of the oceans in winter are noticeably colder than the eastern ones due to frequent intrusions of cold air masses from the continents. In summer, the difference is smaller. Cloudiness over the oceans of temperate latitudes is large and precipitation is significant, especially in sub polar latitudes, where the deepest cyclones are observed.

In the western parts of Eurasia and North America in temperate latitudes, the transport of sea air masses to the mainland both in winter and in summer sharply prevails. Therefore, the climate here bears a strong imprint of oceanic influences and is maritime. It is characterized by not very hot summers and mild winters without a stable snow cover, sufficient precipitation and a more or less even seasonal distribution. This defines the landscape of deciduous forests and meadows. Precipitation increases sharply on the western slopes of the mountains.

In North America, due to the presence of the Cascade and Rocky Mountains, this type of climate is limited to a relatively narrow coastal area. In Western Europe, it spreads inland with a gradual increase in continentally. For example, in Paris (48.8 °N, 2.5 °E) the average temperature in July is +18 °C, and in January - +3 °C; precipitation per year 613 mm. In Europe east of Berlin, the climate becomes continental. Within Russia, this type of climate passes into the continental climate of temperate latitudes. The largest amounts of precipitation fall in this type of climate in Europe on the windward slopes of the mountains. Above, a large amount of precipitation is indicated for Bergen, but at some stations of the Alps, more than 2500 mm falls per year.

In the east of Asia, the climate is typically monsoon. Monsoons of temperate latitudes are a continuation of monsoons of tropical and subtropical, they are very pronounced and are observed approximately up to the latitude of northern Sakhalin. Thus, a moderate monsoon climate is observed in the Primorsky Territory, in northeastern China, in northern Japan and on Sakhalin. In winter, the outskirts of the continent find them on the periphery of the Asian anticyclone, and here the transport of cold air from Eastern Siberia prevails, therefore the winter is low-cloud and dries with significant cold and a sharp minimum of precipitation. In summer, cyclonic activity with fairly abundant precipitation dominates over eastern Asia. An example is the data for Khabarovsk (48.5 °N, 135.0 °E), where the average temperature in July is +21 °C, in January −22 °C and precipitation is 569 mm per year, of which for the winter half-year (October - March) only 99 mm. In the northern regions of Russian Primorye, where the monsoon wind regime is weak or absent, winter is milder due to strong cyclonic activity and the distribution of precipitation is even throughout the year.

On the Atlantic coast of Canada and Newfoundland, the monsoon circulation is weak or absent. Winters are not as cold as in East Asia, and summers are warm enough. In the subarctic (sub Antarctic) belt, arctic (Antarctic) air prevails in winter, and temperate air in summer. In the subarctic belt, two main types of climate are distinguished: continental and oceanic. The continental subarctic climate is observed mainly in the Northern Hemisphere, oceanic - in the Southern.

The tundra zone is located on the northern outskirts of Eurasia and North America in latitudes that can already be called sub polar. Winters are long and harsh, summers are cold and frosty. In Asia, in this type of climate, the cold poles of the Northern Hemisphere (the region of Verkhoyansk and Oymyakon) are located. The average temperature of the warmest month is not higher than +10 ... +12 °C; this is the limit at which trees can grow. Cold summers define the landscape of the tundra. Precipitation in the tundra is less than in the taiga zone, less than 300 mm, and in Eastern Siberia, where cyclones are less likely to get into the tundra zone, even less than 100 mm. Although there is little precipitation, the cloudiness is large and there are many days with precipitation; therefore, they fall out in insignificant quantities due to the low moisture content of the air at low temperatures. The maximum precipitation occurs in the summer. No matter how small the precipitation, but at low temperatures, they exceed the evaporation rate, therefore, excessive moisture is observed in the tundra and swamping due to permafrost.

Over the oceans of the Southern Hemisphere south of 60 ° S lat. to the coast of Antarctica, the subpolar climate is characterized by a very uniform temperature distribution in summer - in most of the water area it is close to 0. But in winter the temperature decreases rapidly and reaches −20 °C and below near the coast.

In these latitudes, cyclone centers often pass, so the cloudiness is very large here; the frequency of precipitation and fog is also high. The prevailing westerly winds near the mainland are replaced by easterly ones.

The climate of the Arctic basin is primarily determined by the lack or complete absence of solar radiation in winter and a very large influx of radiation in summer. The annual radiation balance of the surface of the Arctic seas is generally positive; negative balance only on the Greenland plateau. Nevertheless, summer temperatures are not high, because radiation is spent on melting snow and ice, while the surface and air temperatures remain close to 0. The strong effect of the general atmospheric circulation is added to the influence of radiation conditions. Intensive cyclonic activity is observed in the Arctic basin in all seasons of the year. Cyclones occur on Arctic fronts, and also penetrate the Arctic from lower latitudes, where they develop on polar fronts. Increased pressure prevails throughout the year over the high plateau of Greenland. But in the rest of the Arctic, the division changes significantly from month to month, and anticyclone systems are found on medium maps on relatively small areas in different parts of the ocean, and in winter over Alaska and over northeast Asia. Clouds in the Arctic Basin are generally large and the winds are stronger. Average monthly temperatures in the Arctic Basin are from −40 °C in winter to 0 °C in summer.

The warmest Atlantic-European region of the Arctic, where, in the process of cyclonic activity, deep outflows of warm Atlantic air to high latitudes occur, especially in winter temperatures. In the Asian (East Siberian), Canadian and Pacific regions of the Arctic, winters are much colder than in the Atlantic-European region, but summers are generally the same. At Cape Shalaurova (73.0 °N, 143.3 E) in July +3 °C, in February -31 °C, average annual -15 °C; annual precipitation 140 mm. In the center of the Arctic Basin, average temperatures range from −40 °C in winter to 0 °C in summer.

Greenland, with its high altitude and prevailing anticyclone regime, has a particularly harsh continental climate. On the plateau, at Eismitte station (70.9 °N, 40.6 ° W, 3300 m), the average temperature in July is −14 °C, in January −49 °C, the annual temperature is −32 °C.

The climate of the icy continent is the most severe on the globe. Average annual temperatures decrease here from −10 °C on the coast, at the latitude of the Arctic Circle, to −50 ... −60 °C in the central regions of the mainland. Average annual precipitation for the entire continent is about 120 mm; from the coast to the interior of the mainland, precipitation decreases greatly.

The main role in the severity and dryness of Antarctica's climate is played by the snowy surface of the continent, its high altitude above sea level (on average about 3000 m, and in the center of East Antarctica up to 3500 m and more) and the prevailing anticyclonic circulation regime. Despite the very large summer inflow of solar radiation, the high albedo of the snow surface and effective radiation, the annual radiation balance is negative throughout the continent. The negative radiation balance is replenished by the influx of heat from the atmosphere. Cyclonic activity in the Southern Hemisphere is intensively developing over the ocean around Antarctica. But cyclones penetrate the mainland Antarctica mainly in the western part of the continent. Cyclones very rarely penetrate into East Antarctica. Thus, the high pressure regime is undoubtedly prevailing here. The Antarctic coast is an area with a moderately humid and relatively mild climate. In summer, the maximum temperatures here sometimes rise above zero and the snow melts intensively. Strong katabatic winds from the high plateau of the mainland are especially characteristic of very many coastal areas. Together with cyclones passing near the mainland, katabatic winds lead to the fact that even average wind speeds in a number of places on the coast reach 15-20 m / s. Precipitation on the coast of East Antarctica is 400–500 mm, and that of Western Antarctica is 600–700 mm.

On the glacial slope, in a zone 600-800 km wide (in East Antarctica), strong katabatic winds are also observed, creating a blizzard transport of snow. Average monthly wind speeds are 8-13 m / s. The cloud cover is small, but nevertheless, this zone is often influenced by cyclones passing through the ocean or penetrating deep into the continent. Therefore, snowfalls and blizzards are more frequent here than inside the mainland, and even more often in the lower part of the slope than on the coast. At Pionerskaya station (69.7 ° S, 95.5 °E, 2700 m), the average temperature in August is −48 °C, in December and January −23 °C, the average annual temperature is −38 °C; annual precipitation 848 mm.

On the high inner plateau of East Antarctica, with its predominantly anticyclonic regime, the average wind speeds decrease to 3–4 m / s. Here, over the snow surface, powerful surface inversions persist, and in winter extremely low surface temperatures are observed, up to -90 °C. Average temperatures in winter months are about −70 °C, summer temperatures are about −30 °C. Even in summer, maximum temperatures are below −20 °C. Clear weather prevails; precipitation is very small, about 30-50 mm per year. A third to half of the precipitation is created by hoar frost deposits from ice clouds on the snow cover. The moisture content in the interior of Anthractida is very low.

Under the same latitudes, East Antarctica is much colder than West; in the inland areas by almost 30 °C. This is explained not only by the heights of the area, but also by the fact that West Antarctica is relatively often visited by cyclones that draw warm sea air to the mainland.

The climate of the East Antarctica plateau is characterized by Vostok station (72.1 ° S, 96.6 °E, 3420 m), where the average temperature in August is −68 °C, in December −33 °C, and the average annual temperature is −55 °C, the absolute minimum is −88 °C. Annual precipitation is about 40 mm.

At the South Pole itself, which is closer to the ocean and is more prone to cyclonic activity (Amundsen-Scott station, 2880 m), the climate is somewhat milder: average temperatures in August and September −59 °C, in January −28 °C, during the year −49 °C. Annual precipitation is 55 mm.

Classification of Climates Alisov

Genetic classification of climates, which is based on the division of the earth's surface into climatic zones and regions in accordance with the conditions of general atmospheric circulation, expressed in the prevalence of air masses of a certain geographical type - all year round or in one of two main seasons. The boundaries between the zones are outlined mainly according to the position of climatological fronts in winter and summer. There are 7 main climatic (circulation) zones: equatorial, two tropical, two temperate, arctic and antarctic. Each of them is characterized by a constant predominance of air masses of a geographic type of the same name as the zone. Then, intermediate zones are distinguished: two zones of equatorial monsoons with a winter predominance of tropical and summer equatorial air, two subtropical zones with a winter predominance of polar and summer tropical air, a subarctic zone with a winter predominance of Arctic air and summer air of temperate latitudes. In the tropical and subtropical zones, there are subtypes of climates: continental, oceanic, eastern periphery of oceanic anticyclones, western periphery of oceanic anticyclones; in the temperate zone - subtypes continental, oceanic, western coasts, eastern coasts (monsoon); in the subarctic and arctic zones - continental and oceanic subtypes.

Climate classification based on temperature and precipitation regimes. 5 types of climatic zones are outlined, namely: A - humid tropical zone without winter; B - two dry zones, one in each hemisphere; C - two moderately warm zones without regular snow cover; D - two zones of boreal climate on the continents with pronounced boundaries in winter and summer; E - two polar regions of a snowy climate. The boundaries between the zones are drawn according to certain isotherms of the coldest and warmest months and according to the ratio of the average annual temperature and annual precipitation, taking into account the annual course of precipitation. Within zones of types A, C and D, climates differ with dry winters (w), dry summers (s) and uniformly humid (f). Dry climates, according to the ratio of precipitation and temperature, are divided into steppe climates (BS) and desert climates (BW), polar climates - into a tundra climate (ET) and an eternal (permanent) frost (EF) climate.

Thus, 11 main types of climate are obtained: Af - tropical forest climate, Aw - savannah climate, BS - steppe climate, BW - desert climate, Cw - moderately warm climate with dry winters, Cs - moderately warm climate with dry summers (Mediterranean) , Ci - moderately warm climate with uniform moisture, Dw - moderately cold climate with dry winters, Df - moderately cold climate with uniform moisture, ET - tundra climate, EF - perpetual frost climate.

The genetic classification of climates by B.P. Alisov laid the geographical types of air masses and their circulation. The division of the Earth into climatic zones is associated with the conditions for the formation of climates, which are determined by the circulation of air masses. There are four main climatic zones in each hemisphere:

- Equatorial air,
- Tropical air,
- Air of temperate latitudes,
- Arctic (Antarctic) air.

In each belt, under the influence of land and sea, continental and sea subtypes of air masses and climates are formed. B.P.Alisov's complete system includes the following climates:

Hot Climates

Equatorial Climate

Hot and evenly moist. It is characteristic of the inner Amazon, the Congo basin (Zaire) and the coast of the Gulf of Guinea, the Malacca Peninsula, the Sunda Islands and the island of New Guinea. Radiation balance (residual heat) is one of the main climate-forming processes. On the continents, it ranges from 2510 to 4190 MJ / m 2 (60-100 kcal / cm 2) per year, on the oceans - from 4190 to 5010 MJ / m 2 (100-120 kcal / cm 2) per year. This is less than in tropical latitudes. There are almost no seasonal fluctuations in solar radiation. Much more than residual radiation, the air heats the latent heat of vaporization - up to 4190-5850 MJ / m 2 (100-140 kcal / cm 2) per year. This heat source maintains a uniform temperature flow throughout the year and day.

The circulation of the atmosphere in the equatorial belt consists of two links: 1) convergence of the trade winds and 2) upward currents (vertical convection). The interaction of air masses in the tropical and equatorial zones determines the nature of the moisture turnover: steam comes from two huge tropical zones and condenses in ascending currents in a relatively narrow near-equatorial zone. Naturally, a large amount of precipitation falls here, on average about 2000 mm. The air humidity in the equatorial climate ranges from 81 to 85%, and the annual precipitation is about 2414 mm (Singapore).

Tropical Climate

Typical tropical climate is hot and dry, desert. Variations in the interaction of the "ocean - atmosphere - continent" system determine the pronounced west-east dissymmetry of the tropical nature of the zone. The following tropical climate variations are distinguished:

- Typical desert climates are characteristic of the central continental regions,
- A marine variety of the desert climate characteristic of the western shores of the continents,
- Humid tropical forest and savannah climates inherent in the eastern coasts of the continents.

Solar radiation within the tropical climate is very significant: on the continents 7530-8380 MJ / m 2 per year, on the oceans - 6800 MJ / m 2 (160 kcal / cm 2) per year. The radiation balance on land is lower than on the oceans - 2510 MJ / m 2 (60 kcal / cm 2) per year. This is explained by the fact that 70% or more of the radiation heat is spent on heating the desert sands and then, due to the high albedo of the sands and cloudless sky, is unproductively radiated out of the Earth. The main accumulator of thermal energy - water - is absent here. Deserts are areas of cooling of the atmosphere and geographic envelope.

Anticyclonic atmospheric circulation is characteristic. The descending air masses are adiabatically heated and dried. High atmospheric pressure and radial air spreading prevent the penetration of air masses from the ocean into the western and central sectors. The "ocean-atmosphere-mainland" system does not work for the benefit of these sectors of the mainland. Typical features of the tropical climate are heat, high temperature range and dryness. Large areas are outlined by the July isotherm of 30 °C, sometimes the average July temperature rises to 36.3 ° (Barbara) and even 39 °C (in Death Valley). In winter, the temperature is 10-20 °C. The average annual amplitude is about 20 °C. Much more daily, it can reach 40 °C in the air and 80 °C on the surface of the sand. On the plains, the temperature sometimes drops to -5 °C, and in the Tibesti Mountains to -18 °C. There is not only little precipitation, but they do not fall every year, often they do not happen for several years in a row. The climate of the western outskirts of the continents (a marine variety of the desert climate) is caused by cold currents, reverse air stratification and trade wind circulation, which carries water vapor from the mainland to the ocean.

In the eastern regions of the continents of the tropical zones - on the shores of the Caribbean Sea, in eastern Hindustan, in the Rio de Janeiro region, in Southeast Africa and Eastern Australia - the climate is humid tropical. The sea air masses of the trade winds come here, and a significant amount of precipitation falls. In Rio de Janeiro, the annual rainfall is approximately 1099 mm and it rains every month. The air temperature is higher in Rio de Janeiro in January is 25.8 °C, in July 20.4 °C.

Subequatorial (Subequatorial Monsoon, or Savannah) Climate

The subequatorial climate is, as it were, composed of two independent climatic regimes: equatorial in the July part of the year in the northern hemisphere and tropical in opposite seasons. Therefore, its most characteristic feature is a sharp division into two seasons: the rainy season and the dry season. The air humidity in the subequatorial climate is 70 and 87%; the annual precipitation is 1880 mm. The duration of the dry period in the subequatorial zone ranges from one month at the border with the equatorial climate to 11 months at the border with tropical deserts.

Subtropical Climates

Subtropical climates are between 25 and 40 °N. and S, in the belt of subtropical variable circulation. They are characterized by a seasonal change in the climatic regime: in summer, tropical air with high atmospheric pressure prevails and the weather is dry, and in winter temperate air with low pressure and Central European weather spreads to these latitudes.

The southern (in the northern hemisphere) border of subtropical latitudes is indicated by the southern limit of the winter distribution of moderate air masses or the January position of the moderate front. The northern border of the subtropics runs along the northern limit of the Tula position of the tropical air, or the Tropical High, or, finally, along the July position of the temperate front. The subtropics, therefore, cannot be considered a zone of gradual transition from temperate to tropical climates. These are latitudes with a characteristic seasonal change of temperate and tropical air.

The total solar radiation in the subtropics is 7510-5850 MJ / m2 (180-140 kcal / cm2) per year, that is, it is close to the average for the entire Earth. There is no excess heat here, as in the hot zone, but there is also no negative winter balance, as in the temperate zone. Actual temperatures correspond to solar temperatures.

The circulation of the atmosphere in winter is predominantly cyclopean; relatively cold air masses from temperate latitudes often penetrate. The moisture turnover is characterized by a pronounced seasonality: precipitation falls in winter; summer is dry (except for monsoons).

The greatest longitudinal strike of the continents begins in the subtropical belt of the northern hemisphere; it causes longitudinal differentiation of the climatic zone. Three regions are being formed: western, central and eastern.

The western regions of the subtropical zone are characterized by a Mediterranean climate - Atlas, Southern Europe, Western Asia, areas of San Francisco, Santiago, Cape Town and South-Western Australia, the southern coast of Crimea, the Black Sea coast near the city of Tuapse (Russia).

The inland regions - Eastern Turkey, Iran, southern Central Asia, the Gobi, the southern midwest of the United States, the steppes of Argentina - are characterized by a climate of dry subtropics. It differs from the Mediterranean by the growth of aridity and continentally.

On the eastern outskirts of the continents - In central China and in the Southeast of the United States - the subtropics are humid. The climate of La Plata is close to them. It rains in all months: in winter due to cyclonic air circulation on a moderate front (in monsoon conditions from land), in summer from sea monsoon air.

Moderate Climates

The southern border of the temperate climatic zone runs along the summer limit of the tropical maximum, approximately about 40 °N. and y.sh. The northern border coincides with the winter position of the Arctic Front, approximately with the southern border of the tundra, near the polar circles.

The temperate zone includes Europe, except for the Mediterranean peninsulas, Asia north of the Kara-Bogaz-Gol line - the middle of Honshu Island and up to the tundra zone, North America from the parallel of the Missouri mouth to the latitude of the middle of Hudson Bay. In the southern hemisphere, only the south of South America and the South Island of New Zealand are found in temperate climates. We will restrict ourselves to the characteristics of the northern temperate zone.

The radiation balance in the temperate zone decreases to 1257 and 837 MJ / m 2 (30 and 20 kcal / cm 2) per year. The seasonal difference in the radiation regime is very significant: in summer the balance is about 251 MJ / m 2 (6 kcal / cm 2) per month, slightly less than the tropical one, and in winter it is negative, about 41.9 MJ / m 2 (1 kcal / cm 2) per month. In winter, the temperate zone does not manage with its radiation heat; an important role is played by the advection of heat from tropical latitudes. Heat is transferred by the atmosphere and hydrosphere. Most of it goes to the ocean and peripheral territories, the central regions are cooled.

The main thing in the circulation of the atmosphere is the western transport of air masses, cyclonic and anticyclone activity, and the invasion of the middle latitudes of both Arctic and tropical air. All three air masses - temperate, arctic and tropical - are both continental and marine.

With a drop in radiation heat in the temperate zone, evaporation and the total volume of moisture turnover decrease. Over a large area of \u200b\u200bthe temperate zone, 500-400 mm of precipitation falls, their territorial distribution is characterized by a gradual decrease inland. In the interior regions, which make up most of the belt, snow falls and a stable snow cover is formed that lasts up to six months. The following climates are distinguished in the northern temperate zone:

- Marine, or Western European, or deciduous forests, or, finally, the oak climate. It is characteristic of Western Europe, the coastal strip of Northwest America from Alaska to San Francisco and southern Chile.
- Continental or taiga climate. It is formed in Eurasia from Sweden to the Yenisei, in America - in the forest continental states of Canada and the USA.
- Behind the Yenisei, in Eastern Siberia, the only province on Earth with a sharply continental temperate climate is formed. The advection of heat and moisture from the ocean is the smallest here due to the enormous size of Eurasia. The mainland cools down in winter, a negative temperature anomaly of up to -22 °C forms and a powerful thermal anticyclone is created. There is a cloudless clear winter with little snow with frosts down to -64 °C. A huge area is outlined by isotherms in January at - 30, - 36 and even - 40 °C. Precipitation less than 200 mm; the snow cover is thin, the ground freezes deeply, the permafrost has a thickness of 70 to 800 m.
- An arid temperate climate is formed inside the northern continents in Eurasia to the south, and in America to the west of the forest zone. The climate here is gradually changing towards aridity; forests are replaced by steppes, semi-deserts and deserts.
- The monsoon climate of temperate latitudes, or Far Eastern forests, is typically expressed on the eastern periphery of Asia.

Cold Climates

Cold, or subarctic (sub Antarctic), or tundra, is a climate of variable circulation, in which temperate air prevails in summer and arctic in winter. The cold climate breaks down into two types: mainland and marine.

- Subarctic continental tundra climate is inherent in large areas of the polar margins of Eurasia and North America. The southern boundary of this belt runs along 68 °N, that is, along the 10 °C isotherm, and with the climate of the ice zone - along the 0oC isotherm of the warmest month. The landscape boundaries of the subarctic belt coincide with the boundaries of the tundra. Residual solar radiation in the tundra zone in the annual output is positive, from 209 to 419 MJ / m2 (from 5 to 10 kcal / cm2) per year. But this is only at the expense of two or three summer months; the balance is negative for most of the year. The circulation of the atmosphere, in addition to the already known movements of the Arctic and temperate air also lies in the monsoon tendency: in winter winds from land prevail, in summer winds of northern directions prevail. This lowers the temperature. The thermal regime is maintained by the advection of heat by the atmosphere and hydrosphere, especially in the western regions. The summer is short and cool, the frost-free period is less than 90 days, and the winter is long and cold: January temperatures range from -6 ° on the Murmansk coast to -40 °C at the mouth of the Lena. There is little precipitation: from

300 mm in the west to 100 mm in Eastern Siberia. But evaporation is even less. Excessive atmospheric humidification. The ground is covered with permafrost.

- Oceanic tundra climate is inherent in the islands: Commander and Aleutian in the northern hemisphere, Kerguelen, South Orkney, South Georgia and Falkland in the southern. The oceanic tundra climate on the islands extends up to 500 s. sh. and y.sh. This is due to the cold currents. The climate is even: warm winters, about - 3, - 40 °C, cool summers, below 10 °C. There is a lot of precipitation, 400 mm; large, up to 80-90%, cloudiness in all months; fogs are frequent, the air is damp, it rains many days, and it is almost always windy.

Climates of Perpetual Frost

The Arctic and Antarctic climates are distinguished. The main climatic processes in ice zones are significantly different than in temperate ones. In winter, there is no sunlight at all, and in summer it is around the clock. The snow surface has a high albedo and has a cooling effect on the troposphere.

In summer, within 1-2 months, with the non-setting Sun, the balance is positive, 20.9-41.9 MJ / m 2 (0-1.0 kcal / cm 2) per month. The rest of the 10-11 months the earth only radiates. Deficit of solar heat up to 3350 MJ / m2 (80 kcal / cm 2) per year. With such a radiation balance, the main source of atmospheric heat is the circulation of the atmosphere and hydrosphere. It delivers 3350 - MJ / m 2 (80-100 kcal / cm 2) per year. The air in climates of perpetual frost is usually warmer than the underlying snow surface. In Antarctica, for example, the air temperature near snow drops to -90 °C.

The conditions for heat advection in the Arctic and Antarctic are different. In the Arctic, the troposphere receives from water directly and through sea ice about 167 MJ / m2 (4 kcal / cm 2) per year, Antarctica is deprived of this heat. In summer, the average air temperature in the Arctic is about 0 °C, and on the coasts it reaches 5 °C, in Antarctica - about - 30 °, - 35 °C.

In January in the central Arctic - 40 °C, on the coast of the mainland 30 °C, and on the Atlantic islands, the temperature rises to -16 °C (Spitsbergen). And only over Greenland, a stable anticyclone lowers the temperature in July to -14 °C, and in January to -49 °C.

The climate of Antarctica is the coldest on the entire Earth. In East Antarctica, at Vostok station, the temperature in January (summer) is 32 °C, in August (winter) -71 °C. On the coast it is somewhat softer: in Mirny in January - 2 °. In August - 18 °C.

The described zonal climates are characteristic of lowlands, highlands and low plateaus. In the mountains, climatic conditions change with height, forming a vertical zonation. Climate zones are continuous or discontinuous regions that run parallel to the planet's latitudes. They differ among themselves in terms of air circulation and the amount of solar energy. Topography, proximity to or are also important climate-forming factors.

According to the classification of the Soviet climatologist B.P. Alisov, there are seven main types of the Earth's climate: equatorial, two tropical, two temperate and two polar (one in each hemisphere). In addition, Alisov identified six intermediate belts, three in each hemisphere: two subequatorial, two subtropical, as well as subarctic and subantarctic.

Arctic and Antarctic Climate Zone

Arctic and Antarctic climate zone on the world map. The polar region adjacent to the North Pole is called the Arctic. It includes the territory of the Arctic Ocean, the outskirts and Eurasia. The belt is represented by ice and, which are characterized by long, severe winters. The maximum summer temperature is + 5 °C. Arctic ice affects the Earth's climate as a whole, preventing it from overheating.

The Antarctic belt is located in the very south of the planet. The surrounding islands are also under his influence. There is a pole of cold on the mainland, so winter temperatures average -60 °C. Summer rates do not rise above -20 °C. The territory is located in the arctic desert zone. The mainland is almost completely covered with ice. Land areas are found only in the coastal zone.

Subarctic and Sub Antarctic Climatic Zone

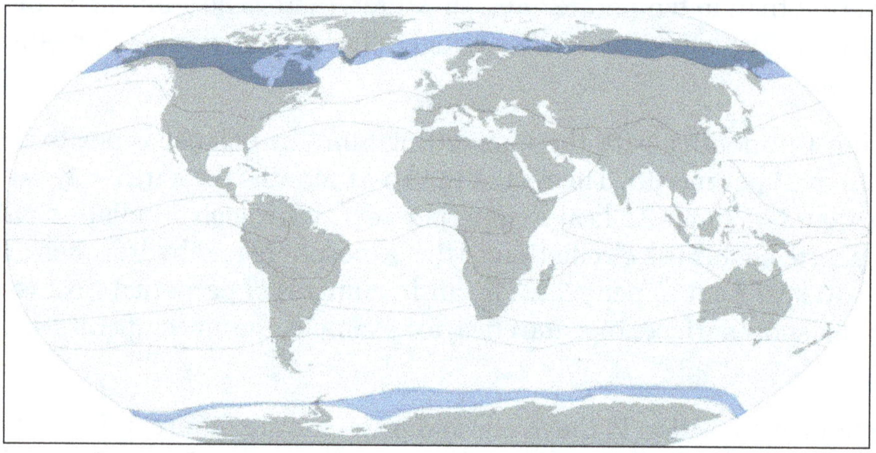

Figure: Subarctic and Subantarctic climate zone on the world map.

The subarctic zone includes Northern Canada, southern Greenland, Alaska, northern Scandinavia, northern regions of Siberia and the Far East. The average winter temperature is -30 °C. With the arrival of a short summer, the mark rises to + 20 °C. In the north of this climatic zone dominates, which is characterized by high air humidity, boggy and frequent winds. The south is located in the forest-tundra zone. The soil has time to warm up over the summer, so shrubs and woodlands grow here.

Within the sub Antarctic belt are the islands of the Southern Ocean near Antarctica. The zone is subject to seasonal influence of air masses. In winter, the arctic air dominates here, and in summer masses come from the temperate zone. The average winter temperature is -15 °C. Storms, fogs and snowfalls often occur on the islands. In the cold season, the entire water area is occupied by ice, but with the onset of summer, they melt. Warm months average -2 °C. The climate can hardly be called favorable. The flora is represented by algae, lichens, mosses and herbs.

Moderate Climatic Zone

In the temperate zone lies a quarter of the entire surface of the planet: North America, and. Its main feature is the clear expression of the seasons of the year. The prevailing air masses give high humidity and low pressure. Average winter temperatures are 0 °C. In the summer, the mark rises above fifteen degrees. Cyclones prevailing in the northern part of the zone provoke snow and rain. Most of the rainfall occurs as summer rain.

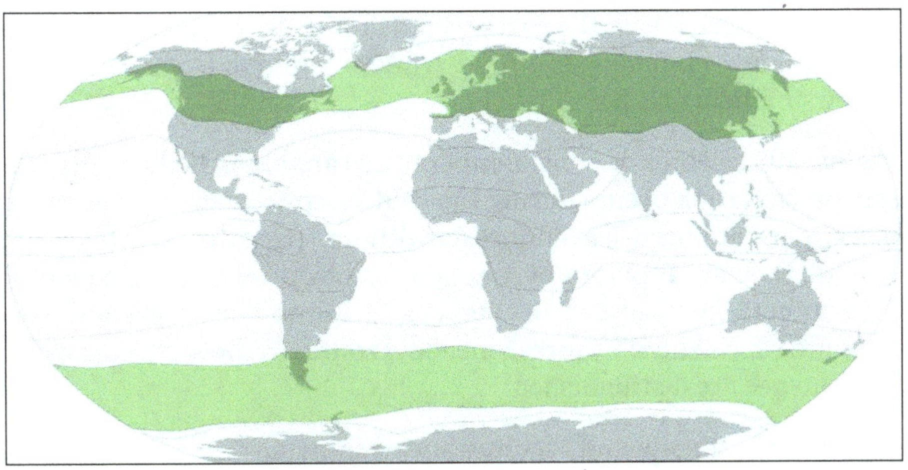

Figure: Temperate climate zone on the world map.

Territories inland are prone to droughts. Represented by alternating forests and arid regions. In the north, it grows, the flora of which is adapted to low temperatures and high humidity. It is gradually replaced by a zone of mixed deciduous forests. A strip of steppes in the south encircles all continents. The semi-desert and desert zone covers western North America and Asia. The temperate climate is subdivided into the following subtypes:

- Nautical,
- Moderate continental,
- Sharply continental,
- Monsoon.

Subtropical Climate Zone

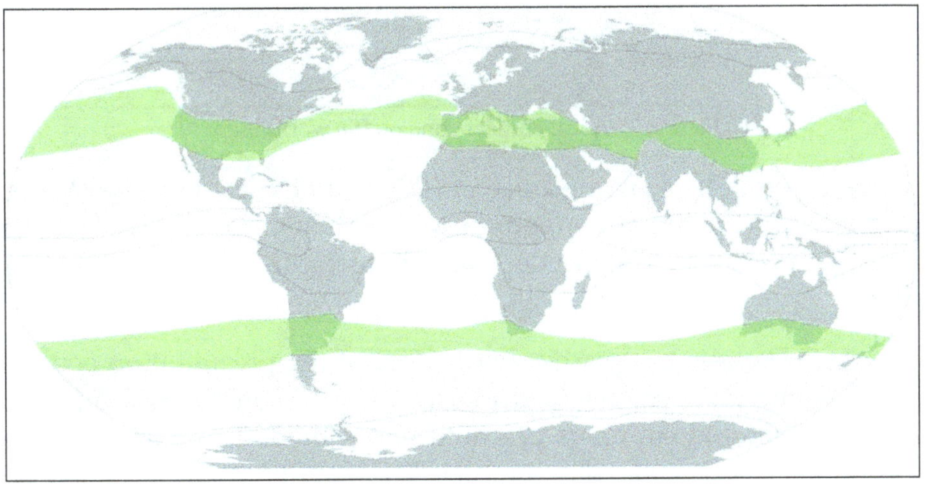

Figure: Subtropical climate zone on the world map.

In the subtropical zone there is a part of the Black Sea coast, southwest and, south of the North and. In winter, the territories are influenced by the air moving from the temperate zone. The mark on the thermometer rarely drops below zero. In summer, the climatic zone is influenced by subtropical cyclones, which warm the earth well. Humid air prevails in the eastern part of the continents. There are long summers and mild winters without frost. The western coasts are characterized by dry summers and warm winters.

In the interior regions of the climatic zone, temperatures are much higher. The weather is almost always clear. Most precipitation occurs during the cold season, when air masses are shifted to the side. On the coasts, there are hard-leaved forests with undergrowth of evergreen shrubs. In the northern hemisphere, they are replaced by a zone of subtropical steppes, smoothly flowing into the desert. In the southern hemisphere, the steppes turn into deciduous and deciduous forests. Mountainous areas are represented by forest-meadow zones. In the subtropical climatic zone, the following climate subtypes are distinguished:

- Subtropical oceanic climate and Mediterranean climate,
- Subtropical inland climate,
- Subtropical monsoon climate,
- The climate of high subtropical highlands.

Tropical Climate Zone

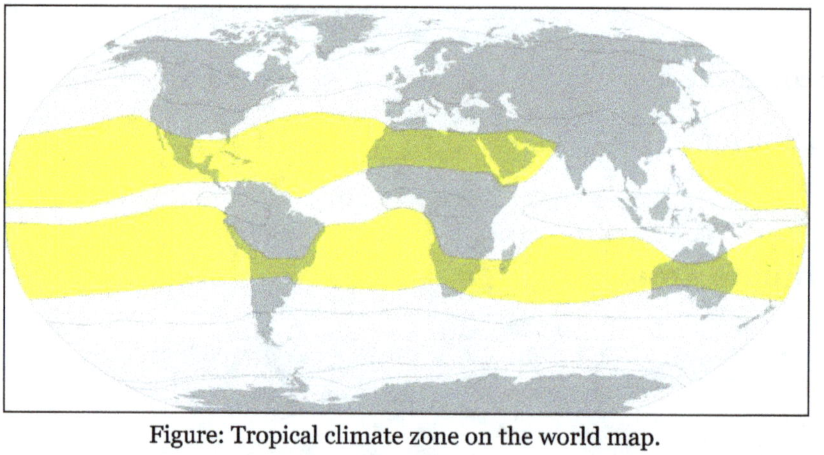

Figure: Tropical climate zone on the world map.

The tropical climate zone covers separate areas in all but Antarctica. Because of this, there is little rainfall in the climatic zone. Summer temperatures in both hemispheres exceed + 35 °C. Average winter temperatures are + 10 °C. The average daily temperature fluctuation is felt in the interior of the continents.

Most of the time there is clear dry weather here. Most of the precipitation occurs in the winter months. Significant temperature changes provoke dust storms. On the coasts, the climate is much milder: winters are warm, and summers are mild and humid. There are practically no strong winds; precipitation falls in the calendar summer. The dominant natural areas are rainforests, deserts and semi-deserts. The tropical climate zone includes the following climate subtypes:

- Trade wind climate,
- Tropical dry climate,

- Tropical monsoon climate,
- Monsoon climate on tropical plateaus.

Subequatorial Climate Zone

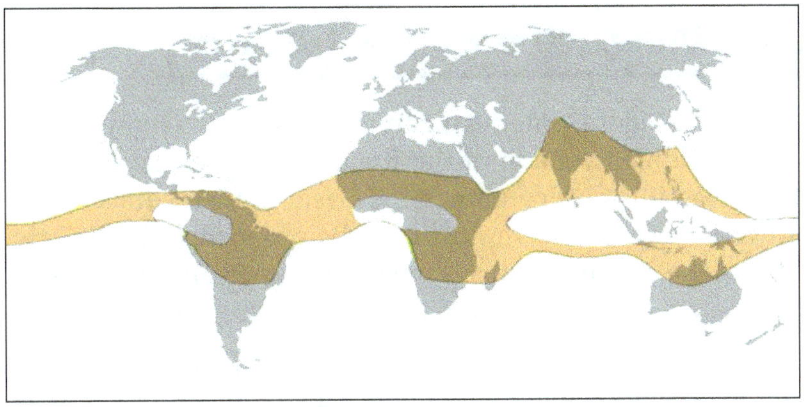

Figure: Subequatorial climate zone on the world map.

The subequatorial climate zone affects both hemispheres of the Earth. During the summer, the zone is influenced by equatorial humid winds. In winter, the trade winds dominate. The average annual temperature is + 28 °C. Daily temperature drops are insignificant. Most of the precipitation occurs during the warm season under the influence of the summer monsoons. The closer to the equator, the more abundant it rains. In the summer, most rivers overflow their banks, and in the winter they dry up completely.

The flora is represented by monsoon mixed forests and woodlands. The foliage on the trees turns yellow and falls off during dry periods. With the arrival of rains, it is restored. Cereals and herbs grow in open areas of savannahs. The flora has adapted to the periods of rains and droughts. Some remote woodland has not yet been studied by humans.

Equatorial Climate Zone

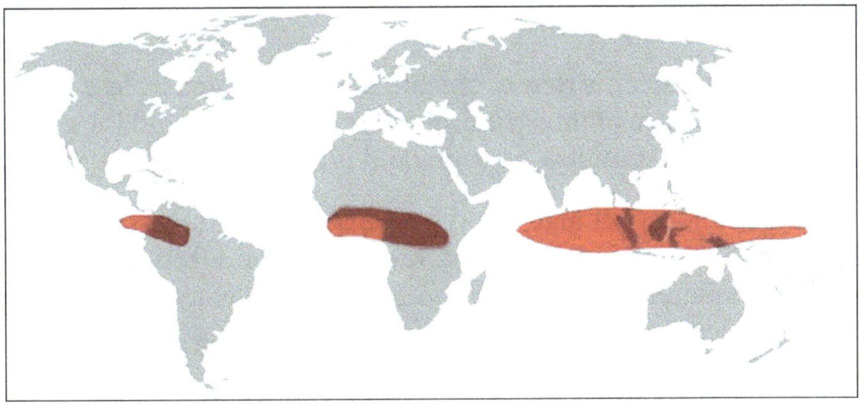

Figure: Equatorial climate zone on the world map.

The belt is located on both sides of the equator. A constant stream of solar radiation creates a hot climate. Weather conditions are affected by air masses coming from the equator. The difference between winter and summer temperatures is only 3 °C. Unlike other climatic zones, the equatorial climate remains virtually unchanged throughout the year. Temperature readings do not drop below + 27 °C. Due to heavy rainfall, high humidity, fog and cloudiness are formed. Strong winds are practically absent, which has a beneficial effect on the flora.

Trewartha's Classification of Climate

Glen T. Trewartha was an eminent American climatologist. The Trewartha climate classification scheme is considered a modified version of the Koeppen system. Besides being 'simple and explanatory, Trewartha's classification combines the basic fundamentals of the empirical as well as the genetic classification schemes. Trewartha, while proposing his climatic classification, was conscious of the fact that the classification systems of Koeppen and 'Thornthwaite being based on certain statistical parameters of certain weather elements were cumbersome and complex.

Trewartha recognised only a limited number of principal climatic types. He made use of the two most important and basic weather elements, i.e., temperature and precipitation as the basic for his own classification.

Besides these, the effects of land and water surfaces on the climate of an area have also been taken into consideration. He has classified world climates in seven climatic groups out of which six—A, C, D, E, F and H—are based on temperature criteria, and the seventh—B—is the dry group based on precipitation.

Climatic Groups based on Temperature Criteria

Group A

This is the tropical climate group. This type of climate is found in the low latitudes on each side of the equator in an irregular belt 20° to 40° wide. There is -no winter season in-this climatic-group. Temperature is uniformly high throughout the year with adequate annual rainfall. In marine areas, the average temperature for the coldest month is around 18 °C to 20 °G. This climatic group is subdivided into two climatic types:

(i) Ar: Ar. is tropical wet climate. This type of climate is characterized by less than two dry months. The climate is under the influence of intertropical convergence zone and the equatorial westerlies. The belt is distinguished by constant low pressure and is also known as the tropical rainforest.

(ii) Aw: Aw is tropical wet-and-dry climate. At the time of the low sun, two months are usually dry. The climatic regions are dominated by the dry trade winds or subtropical anticyclones. During the high sun period, the equatorial westerlies and intertropical convergence control the weather. The duration of the dry season is usually longer than that of the wet season. Temperature remains uniformly high in this type of climate.

Group C

This category encompasses subtropical climate with temperatures above 10 °C for only eight or more months. Frosts occur occasionally in continental parts, but the marine locations remain frostless, On the basis of the seasonal distribution of precipitation, the subtropical climate is further classified into two climatic types:

(i) Cfw: Cfw. is sub-tropical humid climate. This type of climate is found on the eastern side of continents. It has no distinct dry season and rains fall throughout the year. During summer season, this type of climate comes under the influence of unstable air in the western end of a subtropical anticyclone. But & ring winter, the climate is influenced by temperate cyclones.

(ii) Cs: Cs. is sub-tropical dry summer climate. It is characterized by moderate to scanty amount of precipitation. Winter is the rainy season, while summers are nearly or completely dry. This climate type represents a transition zone between the tropical dry climates towards the equator and the temperate climates towards the poles. The average annual precipitation is less than 890 mm (35 inches).

Group D

This group represents temperate climates. The climatic group is also known as the micro-thermal climatic type. The average temperature is around 10 °C for 4 to 8 months. This type of climate is found in the middle- latitudes between the sub-tropical and boreal climates. The two types of climate that are included in the temperate group of humid climates are:

(i) Do: Do. is temperate marine climate. With mild winters, the average temperature for all the 12 months is 0 °C or above. A humid climate with adequate precipitation at all seasons, it is found on the western windward side of the continents in the temperate zone.

(ii) Dc: Dc. is temperate continental climate found in the continental interiors of the middle- latitude continents. The climate is basically land- controlled. The climatic type is characterized by severe winters and summers. Annual temperature ranges are, therefore, high throughout this climate, Cold waves, heat waves, blizzards, and heavy downpours are all yearly events in this category, of climate. Precipitation occurs throughout the year with maximum concentration during summers.

Group E

The group represents sub-arctic or boreal climate found in the higher middle latitudes. Super-continental in temperate features, here the summers are short and comparatively cool. The winters are, however, long and very cold with a very short frost-free season. Average temperature hovers around 10 °C for one to three months during the year. The rest of the year has an average temperature below 10 °C. These regions are characterised by the lowest annual means of temperature for any part of the earth. Even though boreal climates are classified as humid, annual precipitation is comparatively very less. Precipitation occurs throughout the year, most during the warmer months when the amount of water vapour present in the air is highest. Because of the severity of climate, population is sparse.

Group F

The group consists of polar climate found in the high latitudes. The climate is confined to northern hemisphere only. The average temperature in this type of climate seldom exceeds -10 °C. There is no summer season. The polar climates are classified into the following two climatic types:

(i) Ft: Ft. is Tundra climate found only in the northern hemisphere, where it occupies the coastal sides of the Arctic Ocean, and many Arctic islands and ice-free shores of northern Iceland and southern Greenland. No tundra climate is found in the southern hemisphere because of complete absence of extensive land areas.

The Tundra region, essentially a region of grasses, mosses and lichen, is characterised by the absence of trees. The average temperature of the warmest month is recorded between 0 °C and 10 °C. Being a region of permafrost where the subsoil is permanently frozen, precipitation occurs throughout the year, with the warmest season getting the maximum precipitation. In the summer months, precipitation occurs partly as snow and partly as rain. During winters, precipitation occurs in the form of snowfall which is usually dry and powdery.

(ii) Fi: Fi. is ice-cap climate in which the average temperature for all the months is below freezing. There is no vegetation of any kind. The land is permanently covered with ice and snow. The climate is exclusively confined to the ice-caps of Greenland and Antarctica.

Precipitation is meagre, and falls as snow, most of which is in the form of dry, hard, sand-like particles. Most of the snow falls along the seaward margins of the continent. The air close to the ice cap being intensely chilled, surface inversions are the common features of this type of climate.

Group H

The group represents highland climates in which altitude plays a role in determining climate classification. The temperature under normal conditions decreases with altitude, with the summit area of a mountain being always cooler than its base.

Windward slopes force the incoming air to rise up with the resultant condensation, cloud formation and precipitation. The leeward slopes are characterized by descending air which is warmed up and produces little precipitation. East-facing slopes have more sunshine in the mornings which are relatively warmer and cool afternoons. On the contrary, west-facing slopes have cool mornings and warm afternoons.

Trewartha says there is no such thing as a highland type of climate because various types of local climates exist in every significant mountain range. There are no typical temperature and rainfall regimes in the highland climates. They may be considered as the cool, moist islands in the midst of zonal climates that occur in the area around them. The highland areas are also said to be biotic islands with a natural vegetation and animal life adapted to cooler and moister conditions than those of the surrounding low lands.

The highland type of climate are found on the Tibet Plateau, Karakoram Range, the Armenian mountain knot commonly shared by Turkey and Russia, the Alps in Europe, the volcanic highlands of Ethiopia and Ruwenzori Range in Africa, and the Andes in South America.

Climate Group based on Precipitation Criteria

Group B

This group represents dry climate. The boundaries of this type of climatic group are fixed by precipitation values. The characteristic feature of a dry climate is that the loss of moisture through evapotranspiration is far in excess of the annual water gain collected from precipitation. Because of clear and calm weather and the dry atmosphere, the dry climates are quite severe for their latitudes with large annual ranges of temperature.

Precipitation is comparatively very low in this type of climatic group. Relative humidity, high potential evaporation, abundant sunshine and small cloudiness are some of the common features of Group B classification. They are further classified into two climatic groups.

- BW: BW is arid or desert type of climate,
- BS: BS is semi-arid or steppe type of classification.

BW and BS are further classified into the following subdivisions on the basis of temperature:

- BWh: BWh is tropical-subtropical hot deserts,
- BWK: BWK is temperate boreal cold deserts,
- BSh: BSh is tropical-boreal steppes,
- BSk: BSk is temperate-boreal and cold.

The BWh and BWK climates are constantly dry and are under the influence of sub-tropical high and dry trades. The BWh type of condition lasts for 8 months or more with average temperature over 10 °C while the BWK lasts fewer than 8 months with average temperature above 10 °C. The BSh is characterised by a short moist season and is greatly influenced by subtropical high and dry trades. The BSK type of climate receives most of its meagre annual precipitation during the warmer season.

Aridity Index

Aridity is a complex concept that ideally requires a comprehensive assessment of hydro climatological and hydro ecological variables to fully understand anticipated changes. A widely used (offline) impact model to assess projected changes in aridity is the aridity index (AI) (defined as the ratio of potential evaporation to precipitation), summarizing the aridity concept into a single number. Based on the AI, it was shown that aridity will generally increase under conditions of increased CO_2 and associated global warming. However, assessing the same climate model output directly suggests a more nuanced response of aridity to global warming, raising the question if the AI provides a good representation of the complex nature of anticipated aridity changes. By systematically comparing projections of the AI against projections for various hydro climatological and Eco hydrological variables, we show that the AI generally provides a rather poor proxy for projected aridity conditions. Direct climate model output is shown to contradict signals of increasing aridity obtained from the AI in at least half of the global land area with robust change. We further show that part of this discrepancy can be related to the parameterization of potential evaporation. Especially the most commonly used potential evaporation model likely leads to an overestimation of future aridity due to incorrect assumptions under increasing atmospheric CO_2. Our results show that AI-based approaches do not correctly communicate changes projected by the fully coupled climate models. The solution is to directly analyze the model outputs rather than use a separate offline impact model. We thus urge for a direct and joint assessment of climate model output when assessing future aridity changes rather than using simple index-based impact models that use climate model output as input and are potentially subject to significant biases.

The availability of moisture in a natural environment and at climatological time scales is thereby determined through the terrestrial water fluxes (precipitation P, evapotranspiration E and runoff Q), and variables, directly and indirectly, related to these fluxes, such as e.g., atmospheric and soil moisture, and the photosynthetic rate of plants. A comprehensive assessment of aridity changes thus needs to consider the wealth of variables shaping the common perception of a 'dry' landscape. This includes, but is not necessarily limited to, both hydro climatological (e.g. P, Q) and agro ecological drivers (e.g. soil moisture, vegetation cover and productivity, etc.). Assessments of aridity changes under conditions of observed and projected global warming and increased levels of atmospheric CO_2 are due to the complex and somewhat fuzzy definition of aridity are subject to an ongoing scientific debate. A wealth of studies has suggested that global aridity will increase over the 21st century due to the thermodynamic response of a warming atmosphere. The majority of these studies use simple models/indices as a proxy for aridity with the aridity index (AI) approach being the most widely used. Here we follow previous practice and define the AI as the ratio of potential evaporation to precipitation E_p/P (sometimes the inverse, denoted as the wetness index: P/E_p, is used). The AI summarizes the aridity concept into a single number. It is simple and ideally represents the complex interplay of atmospheric and land surface processes that shape the terrestrial dryness. The AI has been found to almost ubiquitously project conditions of increasing aridity under global warming, leading to projections of widespread and accelerated dry land expansion. The AI is further used by the United Nations Environment Programme to classify aridity—from hyper arid and arid to semi-arid. The AI-based framework builds the foundation for the majority of assessments on desertification under climate change—including the most recent version of the World Atlas of Desertification projecting widespread and severe expansions of deserts and semi-arid landscapes. Aridity assessments based on the AI are also featured in various reports of the Intergovernmental Panel on Climate Change (IPCC), including the 5th Assessment Report and the recent Special Report on Climate Change and Land highlighting both uncertainties in aridity projections but also pointing out various severe issues and impacts on natural and socio-economic systems related to potential increases in aridity/desertification.

Estimating aridity based on the AI largely depends on the parameterization of Ep. Since Ep represents a hypothetical flux, it must be modeled from either meteorological observations or climate model output. The most common parameterizations of E_p originate from various models aiming to estimate evaporation (or evapotranspiration) over open water or well watered land surfaces and range from simplistic formulations requiring only limited input variables to sophisticated representations using multiple meteorological and land surface parameters. PMref was introduced as the standard method by the Food and Agriculture Organization (FAO) and has been widely used within the last two decades. The projected increase in PMref is substantial and primarily occurs due to an increase in vapor pressure deficit as a consequence of increasing temperatures and the nonlinear increase of saturation vapor pressure as a function of temperature.

However, considering a distinct vegetation response to elevated CO_2 as simulated in the fully coupled climate models, it is important to point out that some of the assumptions that underlie the computation of PMref (and hence the AI) are incorrect under conditions of changing CO_2 concentrations. A basic assumption in PMref is that the minimum surface resistance over a wet surface is fixed and is thus explicitly assumed to show no response to changing CO_2. This assumption is ultimately not valid for vegetated surfaces as the minimum stomatal resistance—a measure representing the opposition of plants to the exchange of CO_2 and water vapor with the surrounding air—is expected to increase with increasing CO_2, while additionally also soil moisture feedbacks need to be considered. Increases in the stomatal resistance are also expected to lead to an increase in the minimum surface resistance.

Assessments of the impact of using a potentially inaccurate parameterization of the fixed minimum surface resistance in PMref on projections of the AI are missing to date. By considering three alternative approaches that directly and indirectly account for increasing CO_2 concentrations in the parameterization of E_p, our first goal in this study is to investigate potential biases in aridity projections based on the AI calculated using PMref. Comparing projected estimates of AI based on different E_p models—some of them aiming to correct for changing CO_2 concentrations—enables us to assess the impact of using PMref in predictions of future aridity based on the AI. We are further able to assess the sensitivity of aridity projections to the choice of E_p model. The obtained results will thus help to revisit and evaluate previous aridity assessments based on the AI using PMref (projecting widespread and substantial aridity increases) and highlight uncertainties related to different E_p parameterizations.

More generally, a single dimensionless metric such as the AI might also just barely represent the wealth of processes shaping a 'dry' landscape. Similar to the Köppen–Geiger climate classification, the AI potentially provides a sufficient geographical representation of what is usually perceived as arid. However, it is up to debate if projected changes in the AI under conditions of global warming and increasing atmospheric CO_2 provide a good proxy for the complexity of processes driving changes in the mean dryness of the land surface. Hence, instead of using the AI as a secondary, offline impact model (and thereby using climate model output as input) providing a single number to represent the complex aridity concept, a growing number of recent publications have recommended the direct assessment of climate model outputs of relevant Eco hydrological variables. These studies show that, in comparison to the ubiquitous increase in projected aridity as shown by the AI both increases and decreases in aridity occur in different regions worldwide for different variables. A variety of recent studies have also assessed vegetation changes, which are, based on the common definition of aridity, vital to obtain a comprehensive picture of future aridity. These studies show that climate models project a widespread greening under conditions of global warming and increased atmospheric CO_2.

The mixed hydro climatological response and the widespread greening clearly contradict the overall increase of projected aridity in the AI. Hence, by systematically comparing AI projections (ideally representing aridity in terms of a single number) against projections for various hydro climatological and Eco hydrological variables, our second primary goal is to validate if the AI provides a robust proxy for projected aridity changes in general. This will help to evaluate the question if the

AI, a single-numbered impact metric, is appropriate for the communication of projected aridity changes. The validation will hence support a more accurate and comprehensive interpretation of projected changes in aridity. This is especially important given the widespread use of the AI as an impact-relevant metric (for example in the context of desertification).

We aim to obtain a comprehensive picture of anticipated aridity changes as projected by the AI model by comparing in a first step the sensitivity of AI projections to the choice of four different E_p parameterization schemes. In a second step, the AI based on different E_p models is evaluated against both hydroclimatological and ecohydrological variables.

Table: Overview of CMIP5 climate models. We use the following variables from the CMIP5 archive: pr(P), mrsos (uppermost 10 cm soil moisture), gpp (Gross Primary Productivity), and mrros(surface runoff, Q). Required to compute Ep are: hurs(near-surface relative humidity), rsds, rsus, rlds, and rlus(up- and downward long- and shortwave radiation components), sfcWind (surface wind speed), and tas (near-surface air temperature).

Model	AI, P	Q	SM	GPP
ACCESS1-0	X	X	X	
ACCESS1-3	X	X	X	X
CanESM2	X	X	X	
CNRM-CM5	X	X	X	
CSIRO-Mk3-6-0	X	X	X	
GFDL-CM3	X	X	X	
GFDL-ESM2M	X	X	X	X
GISS-E2-H	X	X	X	
GISS-E2-R	X	X	X	X
HadGEM2-CC	X		X	X
HadGEM2-ES	X		X	X
INMCM4	X		X	X
IPSL-CM5A-LR	X	X	X	X
IPSL-CM5A-MR	X	X	X	X
IPSL-CM5B-LR	X	X	X	X
MIROC5	X	X	X	
MIROC-ESM	X	X	X	X
MIROC-ESM-CHEM	X	X	X	X
MRI-ESM1	X	X	X	X
MRI-CGCM3	X	X	X	
NorESM1-M	X	X	X	X

Climate Data

In order to estimate the AI, a variety of meteorological variables is required. These are P and depending on the E_p model: surface net radiation (Rn), temperature, relative humidity and wind speed at 2 m, and atmospheric CO_2 concentrations. We use data from the Coupled Model Inter comparison Project, Version 5 (CMIP5) ensemble. CMIP5 includes climate model projections forced by four representative concentration pathways (RCP) emissions scenarios, corresponding to their relative top of the atmosphere radiative imbalances reached by the end of the 21st century with respect to the preindustrial period. Here we use the high-emission RCP8.5 scenario, consequently resulting in an imbalance of 8.5 W m−2 in the Earth's (top of the atmosphere) radiative budget by 2100. We use a set of 21 climate models, each providing monthly data of the required variables within the time period from 1980 to 2100. The reference period is here defined from 1980 to 1999 by using historical simulations, whereas the future period is set to 2080–2099. All data are regridded to a common 2.5° × 2.5° grid.

The dimensionless aridity estimates obtained from the AI model are compared with simulated hydroclimatological (P and surface runoff, Q) and ecohydrological variables (surface soil moisture and gross primary productivity (GPP), from the same climate models. All data estimates are computed within the same time periods and are regridded to the same common grid. Please note that the surface soil moisture here represents conditions within the uppermost 10 cm of the soil. We argue that a combined assessment of these variables provides a comprehensive (but not necessarily complete) picture of what is commonly perceived as arid. Please note that not all variables are provided by each model (see table for more information on the available variables). It is also important to note that the associated uncertainties in the projected estimates of these variables are substantial. Hence, in the evaluation of the AI against the hydroecological variables, we only consider regions where the majority of models (16 out of 21) agree in sign. Further note that the ensemble of climate models is not weighted, even though several models are from the same modeling center.

Modeling Potential Evaporation

E_p is here parameterized based on four different methods:

Penman–Monteith Reference Evaporation (PMref)

This model is widely-used and recommended by the FAO as the standard method for reference evapotranspiration (FAO-56). It is defined by using a hypothetical crop as reference, assuming a standard plant height of 0.12 m, a fixed surface resistance of 70 sm−1 and an albedo of 0.23, thus ideally representing a surface of well-watered green grass. These values were defined to avoid the need for local calibrations of the aerodynamic and stomatal resistances in the original Penman–Monteith equation, which differ depending on plant and crop type, plant characteristics, and season. To obtain and implement a standard set of resistance terms, bulk formulas have been applied by using the characteristics of the reference crop surface. This implies the use of fixed roughness lengths, a fixed leaf area index of 2.88 and a fixed stomatal resistance of 100 sm−1, resulting in an aerodynamic resistance of r_a = 208/u_2 (with u_2 denoting wind speed at 2m (m s−1)) and a surface resistance of = - r_s 70 sm 1. E_p is thus computed as follows:

$$E_p = \text{PMref}$$

$$= \frac{0.408\Delta(R_n)+\gamma\frac{900}{T+273}u_2(e_s-e_a)}{\Delta+\gamma(1+0.34u_2)}$$

with R_n denoting net radiation (herein MJ m−2 day−1), T mean daily air temperature at 2 m (degC), $e_s - e_a$ the saturation vapour pressure deficit (kPa), Δ the slope of the saturation vapor pressure curve with respect to temperature (kPa/degC), and γ the psychrometric constant (kPa/degC). The term ($0.34u_2$) is finally obtained through the ratio r_s/r_a and represents the suggested reference crop surface. We further estimate $e_s - e_a$ by using relative humidity. PMref can thus be used and applied using standard meteorological output. The FAO further recommends use at daily to monthly time scales.

Corrected Penman–Monteith Reference Evaporation (cPMref)

After establishing a linear relationship between the surface resistance r_s and changes in the atmospheric CO_2 concentrations within a set of 16 CMIP5 models, Yang et al (2019) introduced a modified version of PMref accounting for changing CO_2 levels. By assuming the original = - r_s 70 sm 1 under historical conditions of CO_2 = 300 ppm, they were able to modify the resistance term in PMref depending on the obtained relationship between r_s and the respective CO_2 (units: ppm) levels as (0.34 + 0.00024[CO_2 – 300]). E_p is thus computed as follows:

$$E_p = c\text{PMref}$$

$$= \frac{0.408\Delta(R_n) + \gamma \frac{900}{T+273} u_2 (e_s - e_a)}{\Delta + \gamma\left(1 + u_2\left(0.34 + 0.00024\left[CO_2 - 300\right]\right)\right)}$$

With an atmospheric CO_2 concentration of CO_2 = 300 ppm (representing atmospheric CO_2 levels from the mid-20th century), the modified equation

$$E_p = c\text{PMref}$$

$$= \frac{0.408\Delta(R_n) + \gamma \frac{900}{T+273} u_2 (e_s - e_a)}{\Delta + \gamma\left(1 + u_2\left(0.34 + 0.00024\left[CO_2 - 300\right]\right)\right)}$$

and unmodified equation

$$E_p = \text{PMref}$$

$$= \frac{0.408\Delta(R_n) + \gamma \frac{900}{T+273} u_2 (e_s - e_a)}{\Delta + \gamma(1 + 0.34u_2)}$$

versions are identical.

Net Radiation based E_p

Considering the energy balance at the land surface, net radiationRn provides the maximum estimate of energy available to both sensible and latent heat fluxes. The first method (hereon referred to as Rn80) builds on empirical evidence. Milly and Dunne (2016) found that PMref largely overpredicts changes in evaporation over well-watered surfaces in climate models. However, by using estimates of Rn as the energy constraint to evaporation, they found a reasonable empirical correspondence between wet surface evaporation and ca. $0.8R_n/\lambda$, with λ denoting the latent heat of vaporization (and thus R_n/λ denoting the water equivalent of net radiation). Hence,

$$E_p = Rn80 = 0.8R_n/\lambda$$

However, by further considering the inherent relationship between surface temperature and R_n, being directly coupled through outgoing longwave radiation, Yang and Roderick derived a new formulation (hereon referred to as Rnx) of the maximum evaporation over wet surfaces. This formulation provides the physical foundation to the empirical factor and can be written as follows:

$$E_p = Rnx = \frac{\Delta}{\Delta + 0.24\gamma} R_n \text{ ¢}$$

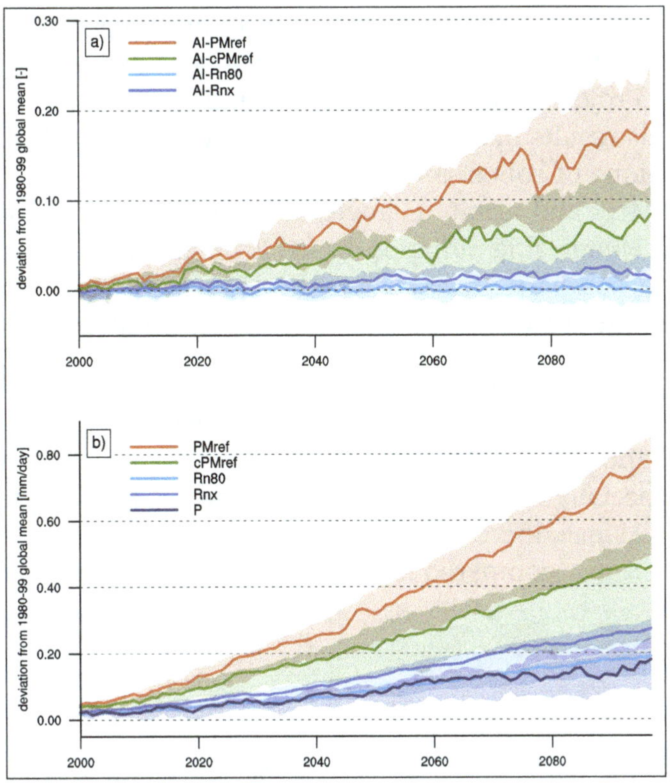

Yearly deviations from the 1980–99 global multi-model mean for(a) aridity index, and (b) P, E_p from 2000 to 2100.(a) AI time series are based on (red) PMref, (green) cPMref, (light blue) Rn80, and (darker blue) Rnx. (b) E_p time series are provided for(red) PMref, (green) cPMref, (light blue) Rn80, and (darker blue) Rnx. Time series of P are provided in black. Solid lines illustrate the multimodel median (21 models considered). The transparent, colored shading represents the respective multi-model interquartile range. A 5 yr running mean was applied to the time series.

Comparing projected estimates of AI based on different E_p models enables us to (i) assess the sensitivity of aridity projections to the choice of E_p model, (ii) evaluate the AI model against hydro-ecological climate model outputs in order to assess whether the AI model provides a good proxy for anticipated aridity changes, and (iii) to assess the impact of changing atmospheric CO_2 concentrations on AI projections.

Sensitivity to Different E_p Models

Using the PMref model, global mean AI is projected to substantially increase over the course of the 21st century. The increase is caused by a larger increase in modeled E_p in comparison to simulated

P. As both Rn80 and Rnx increase at similar rates as P (while the increase in Rnx is, however, slightly larger), there is little change in aridity when estimating E_p through radiation-based models. By correcting PMref for increasing atmospheric CO_2 levels, cPMref increases at a slower rate than PMref. However, the overall increase is still larger than changes in P, leading to an overall increase in AI ranging between the AI time series based on PMref and those time series based on the radiation-based models.

E_p increases in all parts of the world, while the magnitude of changes differs both regionally and between the individual E_p models. The multi-model median changes in E_p between 1980–99 and 2080–99, as illustrated in figure, show the strongest relative increases for PMref, especially in the northern high latitudes. Projected increases in the radiation-based AI estimates are robust, but substantially smaller. This results in stronger projected increases in AI when using PMref, showing large contiguous regions of robust and substantial aridity increases worldwide. This signal weakens when using cPMref and aridity changes are relatively mixed (showing regionally both increases and decreases) when using radiation-based E_p. There are, however, several regions (e.g. the Mediterranean region, southern Africa) where the AI is consistently increasing independent of the choice of E_p parameterization due to consistent and significant decreases in P.

A more detailed illustration of differences between PMref and the other E_p models is provided in figure. In comparison to the projected relative changes in cPMref, changes in PMref are higher by ca. 10%. Such overestimations in PMref are even more pronounced in comparison to the radiation-based estimates, especially for large projected changes in PMref (thus also showing significant conditional biases). Hence, increases in the AI are generally about 20%–30% smaller when using Rnx and especially Rn80, compared to PMref.

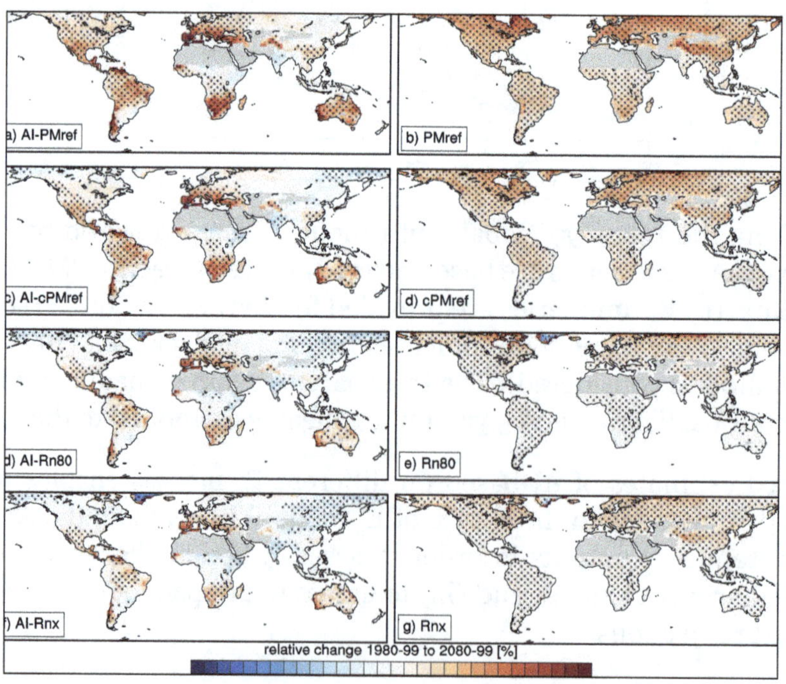

Relative changes in the aridity index (left column) and Ep (right) using the four different methods. Illustrated is the ensemble median change between 1980–99 and 2080–99 under the RCP8.5 emission scenario. Stippling denotes regions where at least 75% of all climate models agree in sign. Grey colors mask regions with ensemble-mean annual rainfall below 100 mm. Please note that based on the definition of the aridity index used here, an increase (decrease) in the index also denotes an increase (decrease) in aridity.

Our analysis thus shows that projected increases in AI are especially pronounced when using PM-ref, whereas such changes are both substantially smaller and regionally more variant for AI based on the radiation-based E_p models. cPMref further clearly reduces the general increases in AI in comparison to the uncorrected PMref model, while projected increase in aridity are, however, still substantial and regionally more pronounced compared to changes in the radiation-based AI.

AI Models a Good Proxy for Anticipated Aridity Changes?

The AI is an impact model ideally providing a reasonable proxy for the historical, current and projected Eco hydrological state of the land surface with respect to long-term dryness in terms of a single, dimensionless number. Regarding projected AI, an evaluation against several Eco hydrological variables can provide a useful assessment of the predictive quality of AI under conditions of global warming and increasing atmospheric CO_2 concentrations. These variables should represent both estimates of land surface water availability, such as P and Q, and estimates representing soil water and vegetation characteristics, such as soil moisture and e.g. GPP, which is here used as an indicator of changing vegetation activity.

The multi-model median changes in P, Q, surface soil moisture, and GPP between 1980–99 to 2080–99 are illustrated in figure. Compared to the respective changes in AI as illustrated in figure, regional increases in P, Q and especially ubiquitous increases in GPP contradict the widespread increasing aridity as projected by AI-PMref in many regions, especially in parts of tropical Africa and East Asia. Apart from decreasing Q (albeit not robust in many regions), changes in P and GPP indicate decreasing aridity in the northern high latitudes of North America, Central, and Northern Europe and Asia, whereas Q is also increasing in parts of Australia. In these regions, AI-PMref and to a lesser extent also AIcPMref, clearly project an overall increase in aridity, thereby contradicting the hydro-ecological response as projected by the fully-coupled climate models.

However, the widespread signal of slight decreases in surface soil moisture partly supports the AI-PMref projections. A better correspondence between projections of AI and the variety of variables shown in figure is indeed found for radiation-based estimates of AI.

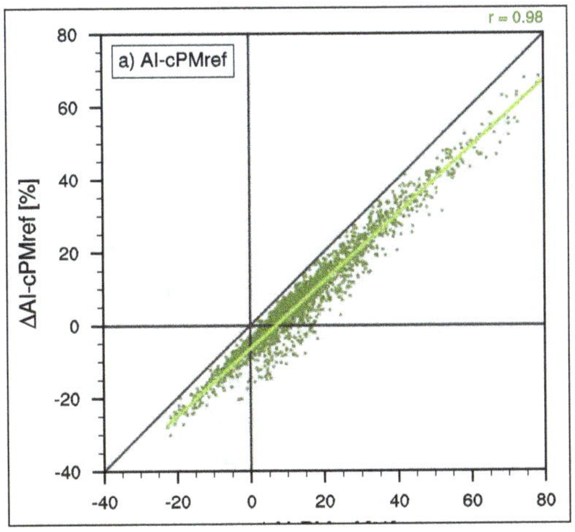

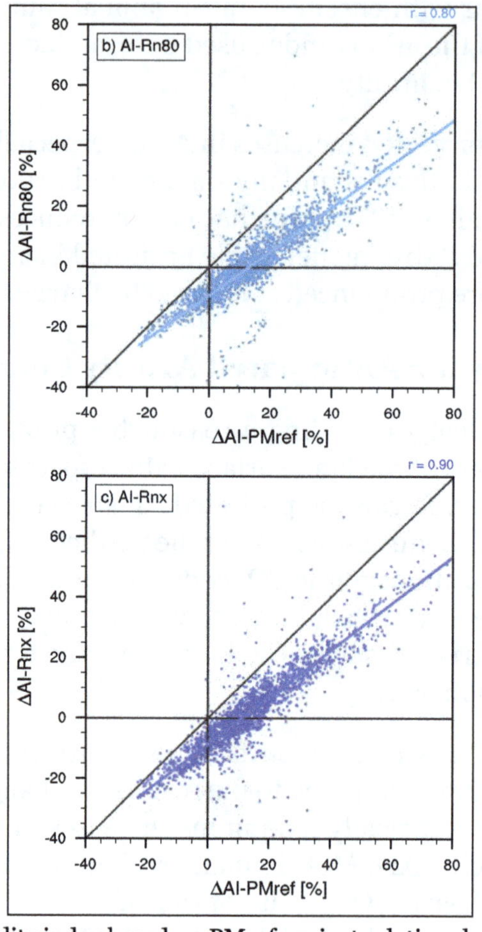

Figure: Relative changes in the aridity index based on PMref against relative changes in the aridity index based on (a) cPMref, (b) Rn80, and (c) Rnx at each grid point. Colored lines illustrate the respective regression lines.

A more detailed evaluation of projected relative changes in AI based on AI-PMref, AI-cPMref, and AIRnx against relative changes in the four hydro climatological and Eco hydrological variables is provided in figure. P (Q) shows robust signals of decreasing aridity in more than 50% (30%) of the land area where robust increases in AI-PMref are projected. Regarding GPP, such increases are found in more than 90% of all grid boxes showing increasing AI-PMref. For P and Q, these percentage numbers are at least halved when using cPMref and even further reduced when using radiation-based estimates of E_p. Such reductions are also substantial for GPP, despite being not as large. Projections of decreasing aridity for the radiation-based estimates and AI-cPMref also correspond better to projected increases in P, Q, and GPP (i.e. also pointing towards decreasing aridity). It can thus be concluded that in only about 44% (65%) of all land area with robust changes (16 out of 21 models agree in sign), aridity increases and decreases in AIPMref and P (Q) correspond. This correspondence is, however, substantially larger (almost 90%) between AI-Rnx and P, Q and almost no correspondence is found between GPP and AI-PMref. However, the AI largely represents projected changes in surface soil moisture independent from the choice of the E_p model, i.e. increases in projected AI almost entirely correspond to regions of decreasing surface soil moisture. Similar to E_p models, surface soil moisture is sensitive to a projected drying of near surface atmospheric conditions and potentially overestimates potential decreases in soil water availability, especially with respect to deeper soil layers that are also more relevant to plants.

References

- Climatology-an-outline, meteorology- climatology, earth-sciences: inflibnet.ac.in, Retrieved 04, May 2020
- Here-are-your-brief-notes-on-koppens-classification-15307: preservearticles.com, Retrieved 08, June 2020
- Useful-notes-on-criticism-of-thornthwaites-1948-classification-15312: preservearticles.com, Retrieved 28, June 2020
- Po-klassifikacii-alisova-razlichayut-zony-klassifikacii-klimatov-vidy-metody, sport-i-fitnes: footyclub.ru, Retrieved 09, May 2020
- Trewarthas-classification-of-climate, climate-818: geographynotes.com, Retrieved 10, April 2020

Chapter 2

Atmospheric Composition and Structure

The gaseous envelope surrounding the Earth is called the atmosphere. It is composed of a mix of several gases in varying amounts, with nitrogen being the most prominent. A study of its composition and structural layers is paramount in understanding atmospheric heating and cooling, insolation, atmospheric circulation, atmospheric moisture, etc. This chapter has been carefully written to provide an extensive understanding of the subject.

The literal meaning of composition is 'ingredients' or 'constituents' of something. In another words, it is a manner by which something is made up of. When we apply the same meaning with atmosphere, it signifies the items or the elements with which our atmosphere is composed. Our atmosphere is composed of numerous gases and other substances; hence, it is a mechanical mixture of the gases, water vapour and dust particles.

Composition of Atmosphere

The envelope of atmosphere around the earth, a mechanical mixture of numerous gases and other substances are very important to all living organisms of the planet. The four major gases – nitrogen, oxygen, argon and carbon dioxide together constitute 99.99% of the total volume of dry air. The maximum concentration is of nitrogen with more than 78 percent while the oxygen is little less than 21 percent.

Groups	Gases	Volume % of dry air
Major Gases	1. Nitrogen	78.084
	2. Oxygen	20.9476
	3. Argon	0.934
	4. *CO_2	0.04
Minor Gases	5. *Methane	0.002
(* are also variable gases)	6. Neon	0.001818
	7. Helium	0.000524
	8. Krypton	0.000114
	9. Hydrogen	0.00005
	10. Xenon	0.0000087
	11. *Ozone	0.00006
Variable Gasses	12. Water vapour	Variable amount
(CO_2, methane and ozone	13. Dust particles	Variable amount
gases are also variable)	14. Aerosols	Variableamount

Nitrogen

Nitrogen is the most abundant found in atmosphere constituting 78.084 percent to the total volume of dry gases. These are almost chemically inactive and have nothing to do with any sort of chemical actions in the atmosphere. It does not combine freely with other elements; hence, it is termed as neutral substance. This gas is found beyond a height of 100 km, but its concentration is below 50 km height from the sea level. This gas is significant for the growth and reproduction in plants and animals. Certain bacteria in the soil are capable of converting a very small amount of atmospheric nitrogen into nitrates and fix it to the soils and water bodies to be consumed by animals and plants. This process is called as Nitrogen fixation. The nitrogen fixed in the earth's surface is again converted and sent back to the atmosphere by bacterial action through a chemical reaction called denitrification.

Oxygen

It is the second largest gas of the atmosphere constituting 20.9476 percent of the total dry atmospheric gases. It is very essential for the survival of many of the living organisms of this planet. It is chemically very active gas. It is combined with several other elements and forms varied compounds. Oxygen is vital for combustion of fuels. When anything burns, oxygen is consumed and helps in burning that substance. Though oxygen is found beyond 100 km but it is reasonably in good proportion within 16 km of height. With increasing height, the amount of oxygen decreases very rapidly. On mountain slope, the available oxygen for breath is very scanty and the mountaineers are supposed to carry oxygen for them.

Gases in Dry Air

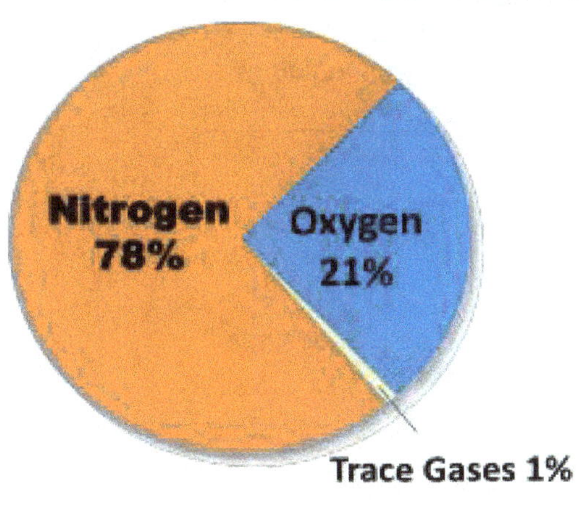

Trace Gases	Percentage by Volume
Argon	0.93
Carbon dioxide	0.038
Neon	0.0018
Helium	0.00052
Methane	0.00015
Krypton	0.00011
Hydrogen	0.00005

Figure: Atmospheric Gases and their Proportions.

Argon

In terms of percentage, argon is the third largest gas in the atmosphere constituting 0.934 percent of total dry atmosphere. It is an inert gas and chemically it is inactive. It is also found in the earth's crust and sea water. It is used in electric bulb and fluorescent lights.

Carbon Dioxide

It is the fourth abundant gas of the atmosphere. It is densest gas and found in lower parts. It is found up to a height of about 30 km; it is concentrated in the lower strata. Its percentage is very low, i.e., 0.04 percent but it is most vital for the growth of vegetative life of biosphere. It is transparent to the incoming solar radiation but does not allow escaping the same. And hence, it is called as greenhouse gas. It plays a very crucial role in increasing the global temperature. It is also known as variable gas as its amount is dependent upon the combustion, human activities and vegetative cover of the planet. The carbon dioxide is reaching to the atmosphere due to several human activities like energy utilization, transport, industry, agriculture, waste generation etc. Apart from these human induced sources, some natural sources are like plant respiration and release to air from stored carbon in the rocks through natural process of denudation. All these are leading to increase in the atmosphere. This gas is on rise with the advancement in economic development of the society. Economically developed countries/ regions are generating big amount of carbon dioxide gases.

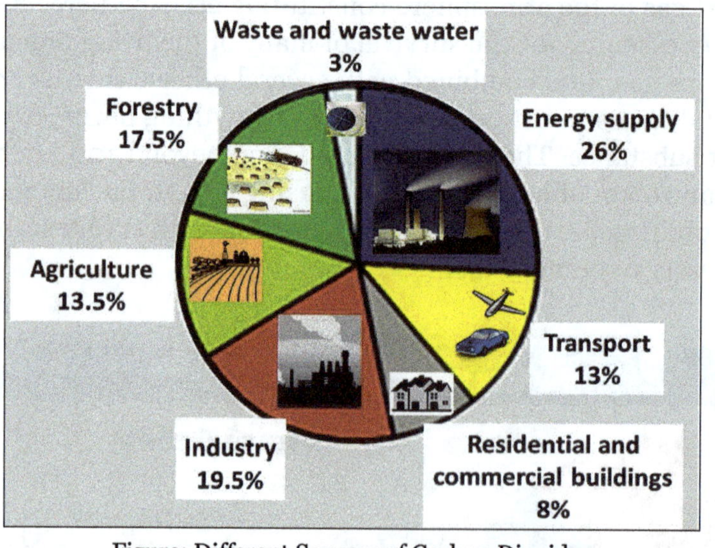

Figure: Different Sources of Carbon Dioxide.

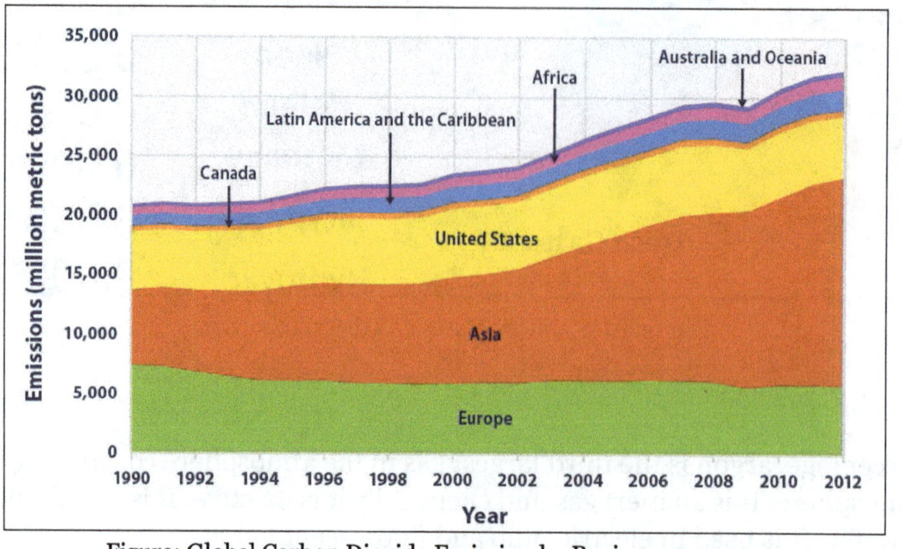

Figure: Global Carbon Dioxide Emission by Region: 1990-2012.

Methane

Methane is the also a greenhouse gas which absorbs the radiation and cause more temperature of the air. Paddy cultivation also generates methane in the air. It is also produced from the wetlands and waterlogged soils and released in the atmosphere. Fossil fuel is also a source to release of methane in the atmosphere. Its amount in the atmosphere is variable. The numerous sources of methane can be seen.

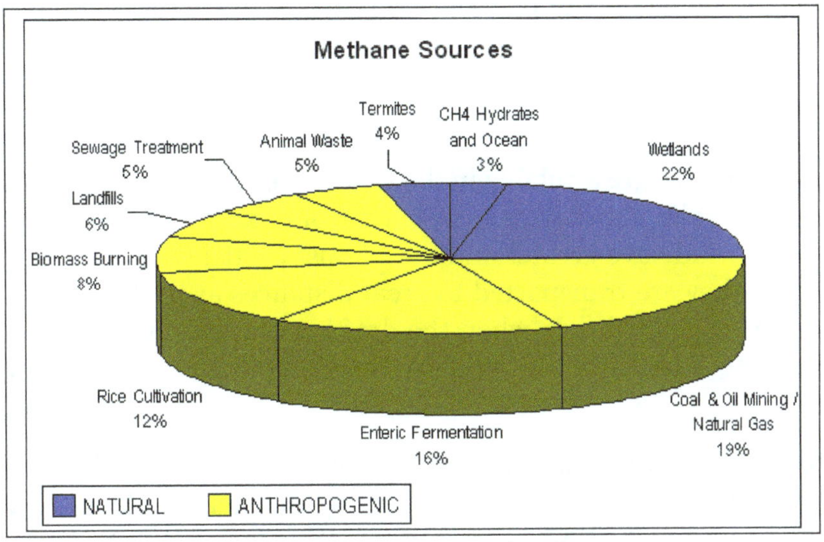

Figure: Different Sources of Methane.

Ozone

Ozone's concentration lies in a belt between the heights of 15 to 50 km of atmosphere. Instead of normal two atoms of oxygen, ozone has three atoms of oxygen formed together denoted by O_3. It is formed when atmospheric oxygen molecules are broken by ultraviolet solar radiation. It may even be formed at the time of electrical discharge during thunderstorms. This gas is also termed as variable as its formation and disintegration is dependent upon numerous activities. Though ozone is very less in quantity (0.00006 percent), this thin layer is very significant for the survival of living world as it absorbs the dangerous ultraviolet rays and protects the earth. Neon, helium, krypton, hydrogen, xenon are other minor gases. Some gases are still extremely less in quantity, they are termed as trace gases. Important among them are ammonia, carbon monoxide, sulphur dioxide, nitrogen dioxide, nitrous oxide and sulphur hexafluoride etc.

Water Vapour

Water vapour is small in amount but it is one of the most important part of atmosphere with respect to the distribution of vegetation and life. Water vapour exists all the time in the atmosphere but with varying degree of amount depending upon the season (temperature condition) and the supply of water for evaporation and evapotranspiration. Air is hardly completely dry. In summer, the water holding capacity of the air is large as the temperature is high while in winter it is low.

Availability of sufficient amount water on the earth surface or water body in an area witness greater vapour while less availability of avoidance of the same shows low vapour. Examples may be taken as equatorial region rich in water bodies (high vapour) and subtropical hot desert region with less to no water availability (low vapour). Though, vapour and air both are in gaseous form, their mixing and movement are quite natural but the same homogeneity it not seen.

At any particular point of time, the amount of vapour is not more than four present of the total volume of atmosphere. It is found in the troposphere only and its concentration is in the lower level. About 90 percent of the total vapour lies below six km. It is estimated that the about 50 percent is within two km of height. Water vapour plays a vital role in keeping the earth warm as it has greenhouse characteristics.

Dust Particles

Huge amount of dust particles are available in the lower layer of atmosphere in a suspended form. These dust particles are solid substances generated from various sources and being carried by winds. Greater velocity wind have greater amount of dust particles. It is not only a matter of suspended solid particles; they are transported to great distances as well. They vary widely in sizes. They may be big sized suspended only when the drafting ability of the wind is great, but settles swiftly when the carrying capacity of the wind is reduced. You must have observed yourselves too during gales/ storms.

Huge numbers of microscopic dust particles are suspended even in completely calm air. Dust particles are variable and it is more during dry seasons as the soils are loose and easily carried by winds, but reverse is the case when it is rainy season when they are settled and compacted. Over the globe, it is less in equatorial and polar areas while more in subtropical hot desert areas. Minute dust particles are found several km above the surface while the coarse sized are abundant near the surface. Microscopic particles are nuclei for condensation and precipitation and they have a great importance in this respect.

Aerosols

Aerosols are extremely fine-sized solid particles or liquid droplets which continue to be in suspended form in gas for very-very long time. They could be seen when their concentration is more otherwise they are invisible. Aerosols themselves are non-gaseous microscopic substance released in the atmosphere from various sources – natural and human created. They could be pollen, minute earthly dust, sea salt, carbon soot from burning fuels, volcanic dust etc. Human activities also help the aerosols to enter the atmosphere. Their concentration is more over the industrial and urban areas. Burning of fossil fuels and generation of smoke also pump the aerosols in the air. Therefore, the sources of aerosols are both natural as well as human generated. They are grouped into two – hygroscopic (moisture absorbing and retaining) and non-hygroscopic (moisture non-absorbing). Hygroscopic aerosols form the nuclei for condensation and in this way, they help in precipitation.

From the above description, it is fairly obvious that our atmosphere is made up of innumerable minute molecules of several gases about which a discussion has been presented above. Apart from that several non-gaseous substances are also available in the airs which are part and parcel of the atmosphere. They have their own significance and play very essential role for the earth to be a livable planet.

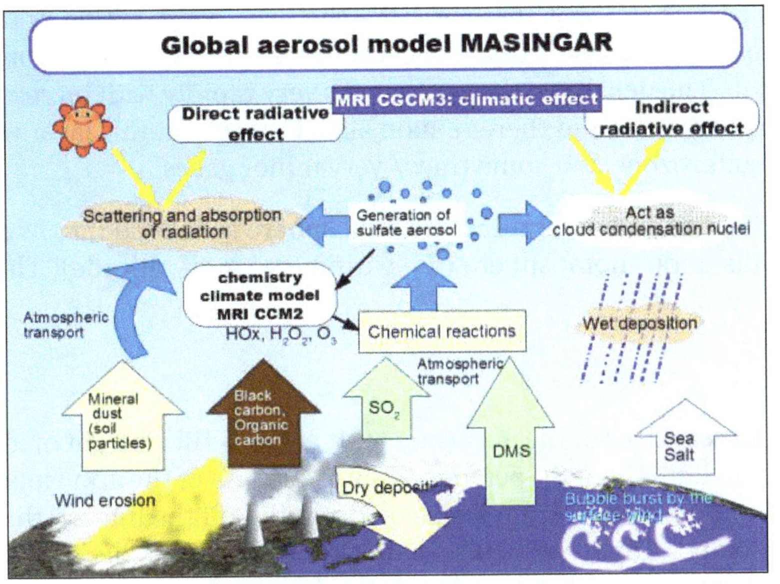

Figure: Different Sources of Aerosols.

Meaning of Structure of Atmosphere

Structure means the arrangement of different part into one. In another words, it is the skeleton or organization or anatomy of a whole by looking at the relationships with its parts. According this background, the study of different parts of the atmosphere and the relationship with its parts is said to be the structure of the atmosphere. Vertically, the atmosphere is divided into different layers/ parts. Therefore, the study of different layers is known as structure of atmosphere.

Structure of Atmosphere

Based on chemical composition, the atmosphere is classified into two. They are homosphere and heterosphere.

Homosphere

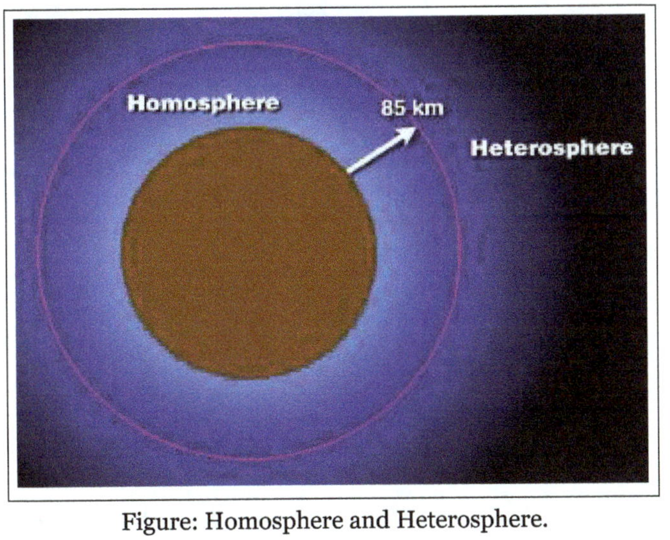

Figure: Homosphere and Heterosphere.

Homosphere is that part of atmosphere where the chemical composition of the air is uniform or similar. It is the lowest layer in terms of chemical composition. It extends from the earth's/ ocean surface to about 85 km. The density of the air changes very rapidly with increasing altitude but the proportion of the major gases found there remain alike throughout this layer with the exception of water vapour, pollutants, ozone and some trace/ very minor gases.

On the basis of the changes in temperature, the atmosphere is divided into five layers out of them, three lower layers falls under homosphere (i.e. within 85 km of altitude). They are troposphere, stratosphere and mesosphere.

Troposphere

It is the lowest and densest layer of the atmosphere. It extends till a height of about 8 km over pole but over equator, it is 18 km. About 80 percent of the total mass of the atmosphere lies in this layer. With increase in height, the temperature keeps on declining till the limit of this layer. On an average, the decrease in temperature with height is 60 Celsius par km. The upper boundary is known as troposphere lying between 8 and 18 km. At this level, the average temperature reaches to minus 50° to minus 60 °Celsius. Water vapour is found in this layer in abundance and about 99 percent of the total atmospheric water vapour is concentrated here but wide variation is seen in terms of height and longitudes. Vapour plays very vital role in regulating the temperature of the earth by creating greenhouse effect.

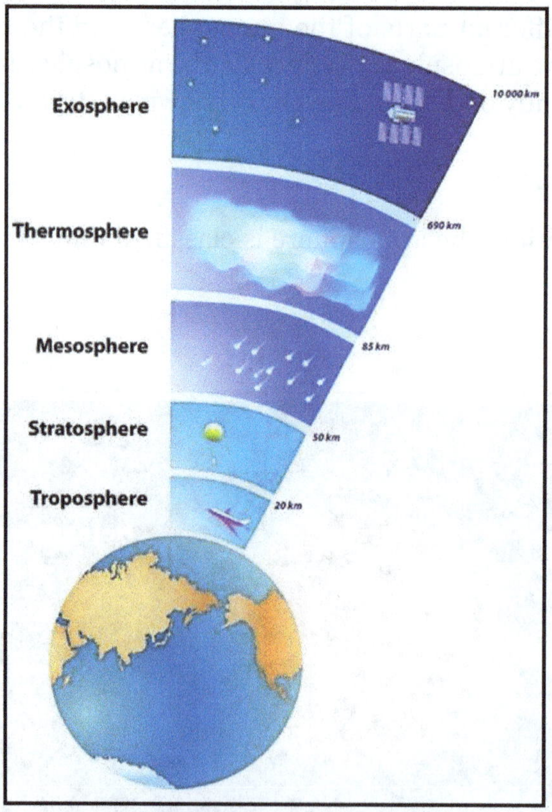

Figure: Different Layers of Atmosphere.

All weather phenomena are occurring in this layer only. Troposphere is the home of all types of clouds, atmospheric turbulence and mixing of the air. Both horizontal and vertical mixing is quite prominent here. In fact, the term troposphere is derived from the Greek word 'tropos' means 'turn'. Sphere is signifying 'ball' or a structure which is round in shape attaining a three-dimensional space. Therefore, the troposphere is a three-dimensional object with turning or mixing characteristics. Every sort of living life is confined to the biospheres which include land water and air. The upper limit of troposphere is tropopause which is a transition zone another upper layer known as stratosphere.

Stratosphere

Stratosphere is the upward second layer as well as middle layer of the homosphere. It starts from tropopause to approximate height of 50 km. The temperature at the tropopause remains almost constant till the height of 20 km. After that, it starts increasing and continues the trend till the height of 50 km. At this level, the estimated temperature is about minus 10° to minus 15° Celsius. Though the temperature is on rise, but there is no atmospheric turbulence. This layer is completely free from clouds and other weather conditions. That is why; it has an advantage for flying long-distance supersonic jets/aero planes through this layer. The increase in temperature in this layer is caused by absorption of solar radiation by ozone (O_3). Ozone is abundant in this layer and its 90 percent is concentrated (found between 15 km to 50 km) in this layer only. The upper limit is stratopause which is a very narrow strip of transition zone beyond which mesosphere is found.

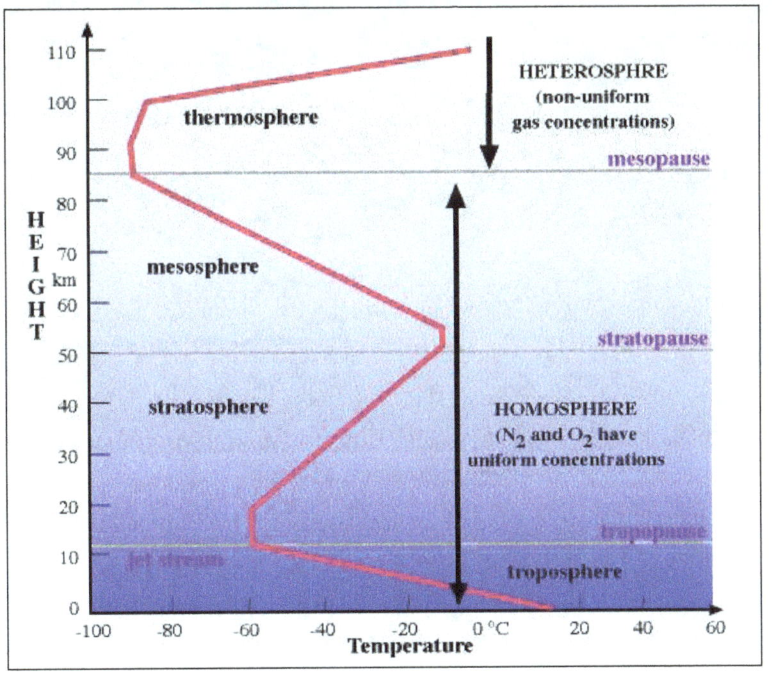

Figure: State of Temperature in Different Layers of Atmosphere.

Stratosphere is also termed as ozonosphere, a layer made up of ozone. This layer protects the living world of the planet from the harmful effects ultra-violet rays. The ozone depletion and hole over Antarctic was discovered in 1985. Since then, we have observed that this layer has reached to a dangerous level of depletion of ozone. The main ozone depleting substances are received from the release of chlorofluorocarbon (CFCs), hydro-chlorofluorocarbon (HCFCs) and carbon tetrachloride. When the molecules of chlorine atom come into contact with ozone (O_3), it turns O3into O_2

which is normal oxygen. One chlorine atom can destroy more than a lakh atom of ozone. Therefore, the chain of depletion is very serious. O_3 turned into O_2, an ordinary oxygen molecule is not capable to absorbing ultra-violet rays.

Mesosphere

Mesosphere is the third but the upper-most layer of the homosphere. After this layer, heterosphere starts. The literal meaning of mesosphere is the middle sphere. It is separated by tropopause below from troposphere and mesopause on the top from thermosphere. It is extended from 50 km to 85 km from the earth's surface. The air pressure is very low. It is 1 millibar at the lower limit whereas it is 0.01 millibar at the highest limit. This layer is characterized by decreasing temperature and the coldest/ lowest atmospheric temperature is recorded in this layer. The lowest temperature estimated near the mesosphere is around minus 130 °Celsius. It is colder that the lowest temperature recorded over Antarctic.

Between 75 to 85 km from the earth, noctilucent clouds are normal affairs in the summer nights between 50° to 70° north and south latitudes. Its literal meaning is night shining. It is a deep twilight seen only when the sun is on horizon but the sunlight is still falling at that height. The seen clouds are made up of ice crystals. Meteoric dust particles work like nuclei for ice crystallization which are falling as well as produced due to burning of meteor caused by friction.

Figure: Noctilucent Clouds.

Heterosphere

The atmosphere lying beyond the homosphere is termed as heterosphere. The term itself is self-explanatory and it is used for that part of atmosphere where the air is not uniform. In this part of atmosphere, the air is rare and the molecules are wide apart. Relatively heavier gas molecules are concentrated in the lower part whereas the lighter are forced to be above.

Beyond 85 km height, the compositions of the atmosphere with increasing altitude vary significantly. The mixing of the gases is not possible as the turbulence is not happening there. Different layers of prominently different gases are nitrogen layer; oxygen layer, helium layer and hydrogen layer are differentiated. However, the heterosphere, is divided into two main spheres – thermosphere and exosphere.

Thermosphere

This sphere extends from mesopouse i.e., 85 km to about 650 km from earth. The temperature is on rise in this layer due to absorption of solar radiation by small amount of oxygen molecules present. It is highly dependent upon the solar activities. The temperature reaches beyond 1200 °C at an altitude of about 350 km but by 650 km it may even rise to 2000 °C. This much high temperature is primarily defined by average speed with which molecules are moving. Because of this, the temperature may be high. The effectiveness of this temperature is not that great. Its exposure to astronaut, if they are coming out from the capsule, is not affecting at all.

The Aurora is conspicuous phenomena observed in this layer. It is a striking display of light and is maximum between 10pm to 2am in the magnetic polar region of northern and southern hemispheres. It is known as Aurora Borealis in the northern and Aurora Australis in southern hemisphere. They are referred as northern and southern lights. Charged particles' coming from the solar flare reaches to the earth's upper atmosphere. These charged particles collide with nitrogen and oxygen molecules present there in rarity. The collusion results into innumerable little surges of lights. The scattering of these lights seen is very beautiful and magnificent.

Figure: Aurora Borealis over Alaska.

Exosphere

Exo means external. Therefore, exosphere the external or the outer most layer of the atmosphere. Its lower boundary starts from the thermopause (650 km) to the limit from where the void space begins. This limit is estimated to be about 10000 km. This much distance is little less than the diameter of the earth. It is really a very big size of the limit of the atmosphere. In exosphere, very light gases are traced and they are hydrogen and helium. Their molecules are spaced very widely. Beyond the upper limit of exosphere, the space is considered to be void.

Significance of the Atmosphere

Earth's atmosphere is a thin blanket of gases and tiny particles: together called air. We are most aware of air when it moves and creates wind. All living things need some of the gases in air for life support. Without an atmosphere, Earth would likely be just another lifeless rock. Earth's atmospheres, along with the abundant liquid water at Earth's surface, are the keys to our planet's unique place in the solar system. Much of what makes Earth exceptional depends on the atmosphere.

Indispensable for Life of Earth

Without the atmosphere, Earth would look a lot more like the Moon. Atmospheric gases, especially carbon dioxide (CO_2) and oxygen (O_2), are extremely important for living organisms. How does the atmosphere make life possible? How does life alter the atmosphere?

In photosynthesis plants use CO_2 and create O_2. Photosynthesis is responsible for nearly all of the oxygen currently found in the atmosphere. By creating oxygen and food, plants have made an environment that is favorable for animals. In respiration, animals use oxygen to convert sugar into food energy they can use. Plants also go through respiration and consume some of the sugars they produce.

Crucial Part of the Water Cycle

As part of the hydrologic cycle, water spends a lot of time in the atmosphere, mostly as water vapor.All weather takes place in the atmosphere, virtually all of it in the lower atmosphere. Weather describes what the atmosphere is like at a specific time and place, and may include temperature, wind, and precipitation. Weather is the change we experience from day to day. Climate is the long-term average of weather in a particular spot. Although the weather for a particular winter day in Tucson, Arizona, may include snow, the climate of Tucson is generally warm and dry.

Ozone Layer makes Life Possible

Ozone is a molecule composed of three oxygen atoms, (O_3). Ozone in the upper atmosphere absorbs high-energy ultraviolet (UV) radiation coming from the Sun. This protects living things on Earth's surface from the Sun's most harmful rays. Without ozone for protection, only the simplest life forms would be able to live on Earth.

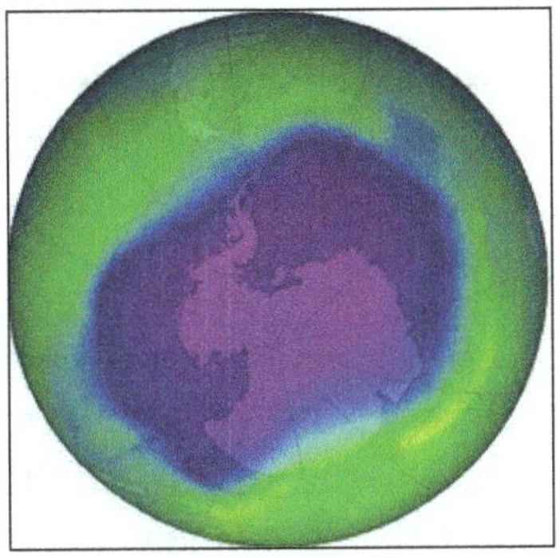

Moderates Earth's Temperature

Along with the oceans, the atmosphere keeps Earth's temperatures within an acceptable range. Greenhouse gases trap heat in the atmosphere so they help to moderate global temperatures. Without an atmosphere with greenhouse gases, Earth's temperatures would be frigid at night and scorching during the day. Important greenhouse gases include carbon dioxide, methane, water vapor, and ozone.

Atmospheric Heating and Cooling

The source of heat on the earth and for earth's atmosphere is almost completely governed by the energy received from the sun. Little heat reaches to the surface and atmosphere by earth's cooling, hot springs, and volcanic eruption. Volcanic eruption is confined to certain belts/ pockets on the globe. Primarily these sources are very meager keeping the size of whole of the earth as well as huge mass of surrounding atmosphere. It is the solar energy which is responsible for the distribution of heat/ energy all over the globe. The distribution of this energy is again affected by numerous factors.

Mechanisms of Heating and Cooling

Heating and cooling of the atmosphere is performed by following processes:

- Partial absorption of solar radiation by atmosphere,

- Conduction,
- Terrestrial radiation,
- Convection,
- Advection,
- Latent heat of condensation,
- Expansion and compression of the air.

Partial Absorption of Solar Radiation by Atmosphere

The solar radiations are coming to the earth surface directly from the sun. It is in the form of short-wave radiation. They are so energized that atmospheric gases are unable to trap them. But the presence of some dust particles and water vapour in the lower level of troposphere are capable of holding some energy directly coming from the sun.

About 20% of the total incoming solar energy is trapped by the dust particles and vapour in the atmosphere. Approximately, half of the total absorbed energy is done so within two km of the height from the earth's surface. It is this layer where the amount of dust particles, fire shoots/ smokes and water vapour are available in large quantities. However, the quantity of these particles decreases sharply with increasing height.

You must have also observed that when the patches of clouds are there between you and the sun in day time, you feel cool even when the temperature is very high. It is so, because the concentrated water vapour (clouds) acts as an intervening obstruction by holding some part of solar energy between sun and the earth. In this process, the air is heated where it contains vapour, dust particles and fire shoots. But the heating of air over certain height through these obstructions does not increase the surface temperature in that particular region. Since, these areas are receiving relatively less energy on the surface, the surface is cool as otherwise it would have been. Because of more cloud cover in equatorial regions, relatively less energy is received at the earth surface in these regions. On the other hand, the sky is clear over subtropical belt, hence the energy received in the surface areas of these belts is quite high.

Conduction

The literal meaning of the term conduction is passing on something by a medium without any perceptible movement by itself. It is the transfer of something from one part to the other without any physical movement.

You must have observed, while vegetable is cooked in a pan and a metallic spatula is used to turn the vegetable, it gets heated up while its handle is outside the pan. It happens so, because the heat received by the pan from the burner and the spatula is touching the pan. The received heat of the pan is transmitted to spatula which is warmed up and thus it burns our hand if left for a little more time in the pan.

Air is a very poor medium of heat conduction. It is a very slow process of transferring heat in a mass of air. By this method, air is heated, but its importance is not that great. Because, the air is in the gaseous state and its particles (molecules or atom) are not very solidly compacted. A very thin layer which is very close to the earth surface is heated by conduction method. Once the air atom is heated, it becomes lighter and less dense and ultimately moves upward. Therefore, conduction has a very insignificant role in heating the atmosphere. It is almost negligible in comparison to other methods. Practically speaking, majority of the meteorologists and climatologists prefer to neglect the conduction method of atmospheric heating.

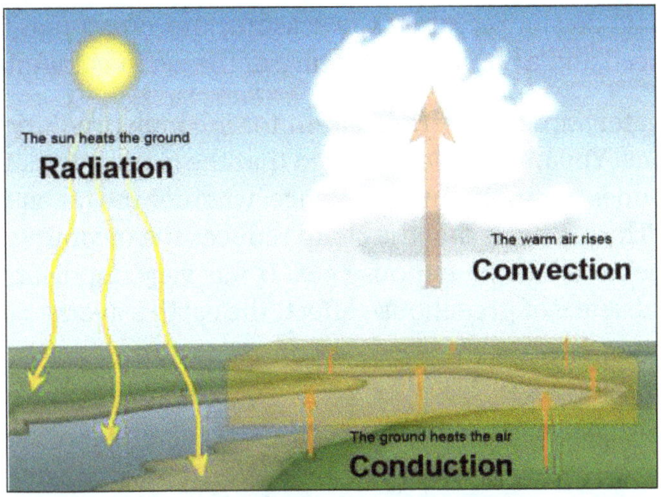

Figure: Atmospheric Heating Methods.

Terrestrial Radiation The terrestrial radiation is the most important method of atmospheric heating. Out of the total solar electromagnetic radiations reaching at the top layer of the atmosphere, approximately 49% reaches to the earth's surface. Out of this, 5% is reflected back to the space without heating the atmosphere. About 20% is absorbed by the atmospheric substances including water vapour. Therefore, 45% (49-5=45%) plus 20% (total 65%) is available for heating the atmosphere.

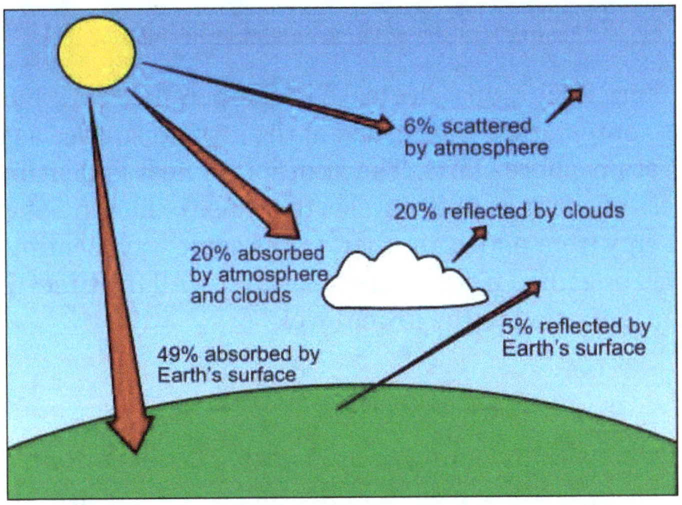

Figure: Solar Energy at the Earth and in its Atmosphere.

All the above mentioned energy is reaching the earth surface in the form of short wave electromagnetic radiations from the sun. The heated earth radiates back the same but in the form of long wave electromagnetic or infrared radiations. The atmosphere gains the heat radiated through long waves from the earth surface. Most of the short wave radiations are not being trapped by the atmospheric gases, as they are not capable of.

Terrestrial radiations are a continuous affair all 24 hours throughout the year. During day time when sun is in the sky, the solar short wave incoming energy is greater than the energy lost from the earth surface (land and water). Since there in no incoming solar energy during nights, the day time received energy is radiated back to the atmosphere. There may be addition or subtraction of energy on day to day basis depending upon the seasons, but on annual basis, the incoming and outgoing energy is balanced and a static temperature of the earth is maintained.

Greenhouse gases and water vapour are transparent for incoming radiation but they trap the outgoing long wave radiations. You must have observed that the cloudy day keeps us cool while cloudy night makes us warm. It does not make any difference, whether the taken examples are of summer or winter day and night. The reason is that the cloud reduces the incoming radiations and traps the outgoing radiations. It behaves like greenhouse gas. It has very important role to make the earth livable otherwise in the absence of greenhouse effect, the earth's average temperature would have gone down to minus 17 °C.

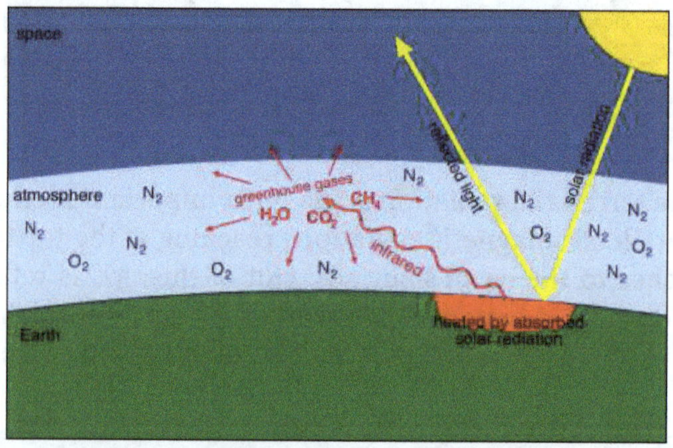

Figure: Heating of Earth and its Atmosphere.

It is also very true that there is a gradual decrease in temperature with increasing altitude within the troposphere. Greater temperature is recorded at the ground surface as the earth is heated first and then the heating of atmosphere starts. The addition of heat is distributed and finally results into a cooling. The process of cooling at one level is the cause of heating at another level. It is nothing but the transfer of energy from one to another. In the same way, heating is possible because the energy is again distributed or gained from where it has more. All the time, the energy is transferred from the high level of concentration to the lower level.

Convection

The earth's surface is heated with incoming solar energy. The air in contact with the surface is in gaseous form. Earth's surface heating results in heating of the air in its contact. But the air becomes less dense by heating. It further results into rising of the warmed/ expanded air molecules upward. Upward moving air molecules in large quantity create convection. The removed air by expansion at the lower/ ground level results into creation of low pressure. Therefore, nearby air from relatively cool air starts moving to fill the gap created by upward lifting of air. The heat is also transferred upward with vertically moving air. This convection may occur at a local level as well as at much larger regional level. The occurrence of Hadley cell, Ferrel's cell and Polar cell are examples of atmospheric convection. Therefore, convection transfers the heat energy received from the sun to the surface and from the surface to the atmosphere. The Figure very clearly explains this.

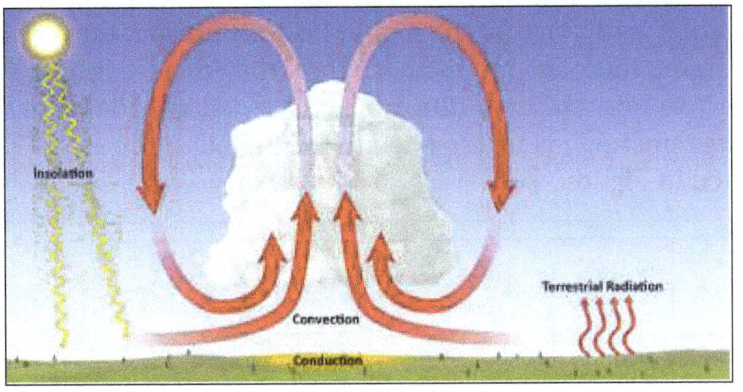

Figure: Convectional Heat Transfer in the Atmosphere.

Advection

The meaning of advection is transfer of something from one place to another especially in horizontal direction. Atmosphere is a huge body of air and it has differences in terms of its pressure depending on several affecting factors. Due to varying pressure at local, regional and global level, atmospheric gases are continuously on move. A local level horizontal movement of air (advection) can be seen from Figure. The monsoonal air current movement is the example of regional advection while planetary permanent wind system signifies the global advection. All of them are transferring the heat from one area to another. Transfer of energy is not only done by atmospheric advection, but also by hydrospheric advection. There is movement of water also from the low latitude to high latitude in the form of ocean currents. The ocean currents also redistribute the received energy from high concentration to low concentration zones. It is also a fact that the ocean currents are also affected by the adjective movements in the atmosphere.

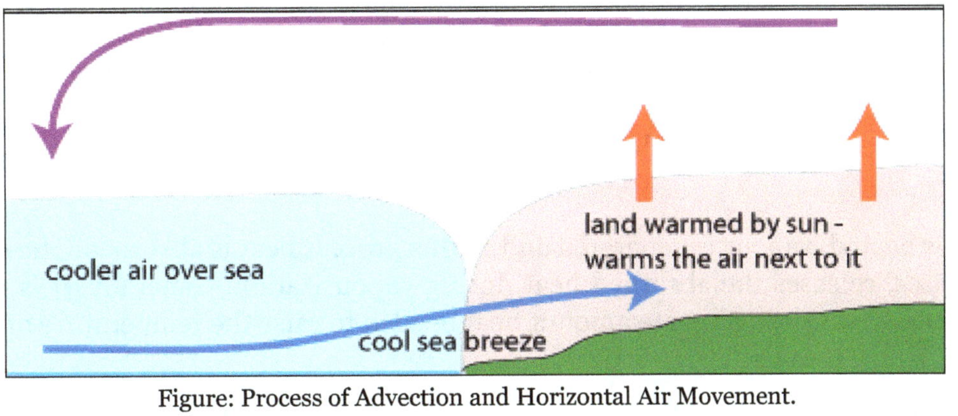

Figure: Process of Advection and Horizontal Air Movement.

Once the air is on move horizontally, physical transfer of air, associated heat and other related properties of the mass of air are transferred. In this process, the heating of the atmosphere is the result. It is caused by the differences in air pressure and temperature. In fact, these two characteristics of air are primary factors in the movement of the mass of the air.

If the net radiation at global level is calculated for the atmosphere, it is established facts that there is positive budget between incoming and outgoing energy between 40° north and 35° south latitudes. This fact is very clear from the Figure. It is also apparent that there is negative budget beyond 400 north and 35° south latitudes. Overall the total budget of the earth is almost neutral. The positive received radiation is diverted to the higher latitude areas by advection. Therefore, higher latitude areas are releasing more energy to space than they receive as it is transported from low latitude areas. Contrary to this, the low latitude areas are releasing less energy than they directly receive from the sun as remaining is transported to the high latitude areas. The same information is very clearly presented in Figure.

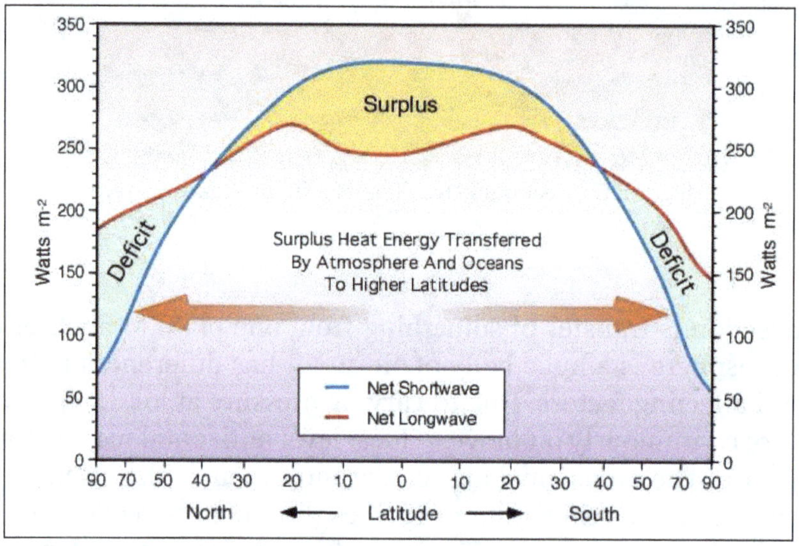

Figure: Surplus and Deficit Zones of Radiations.

Latent Heat

Heat absorbed or released due to change of the state of any matter is known as latent heat. During this process, there is no change of temperature of that matter. In another words, it is the heat that is required to change the matter to a higher state of matter. You may be well aware, when water changes from one state to another, for example, water vapour to liquid water and liquid water to solid water (ice), it absorbs or releases heat. The energy involved in this process is known as latent heat, popularly meant for 'hidden' heat. Water vapour transports the energy from one region to another.

When water is heated, vapour is generated and in this process heat is absorbed. The same vapour when condenses, releases the absorbed heat during vapourization. Generally, it is measured in calories. One calorie is the aggregate amount heat needed to raise the temperature of one gram of water by one degree Celsius.

Suppose one gram of ice is melted, it releases 80 calories of heat and turns into water. If the same one gram of water is evaporated, it requires 540 calories. Therefore, the said calories are known as latent heat. In the same way, when the condensation is taking place, conversion of vapour into one gram water releases 540 calories and it reaches to the atmosphere and heats it. When one gram water turns into ice, 80 calories is released what it was consumed while melting the ice. Sublimation is the process in which ice turns into vapour (it releases 540+80=620 calories) or vapour turns into ice, (it releases 540+80=620 calories). In both the cases, latent heat is either released or utilized Figure.

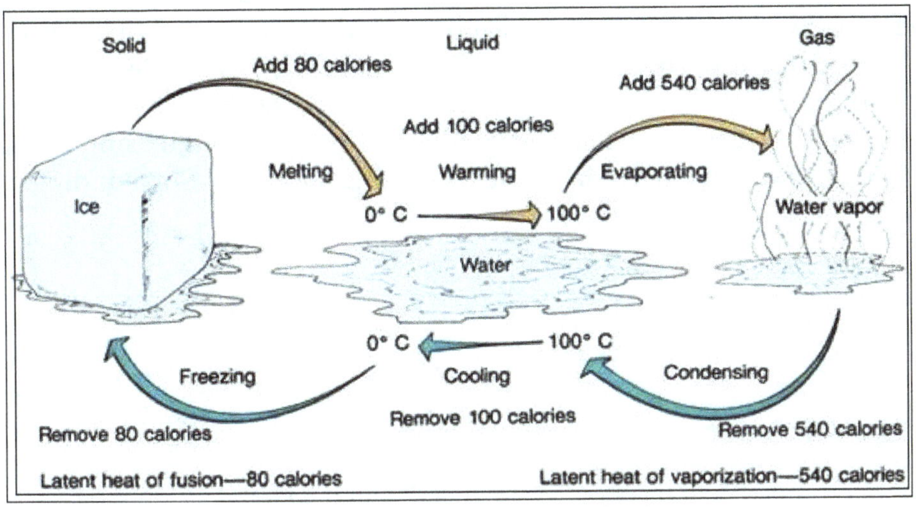

Figure: Latent Heat: Absorption and Release.

It is quite apparent that the latent heat plays a significant role in heating and cooling of the atmosphere. When the heat is utilized in turning the state of matter from lower level of heat to higher level, cooling is evident. But when it is in descending order, i.e., from higher level of heat to lower level, it releases the latent heat and heating of the surrounding atmosphere is the result.

You may even feel cool when the sweat is removed from your skin either by moving fan or the use of air conditioner. Removal of sweat by transpiration/ evaporation is associated with latent heat, and hence we feel comfortable. When condensation takes place, it leads to release of the same latent heat and causes the air to warm up at the level (generally upper level of atmosphere) of condensation. The same thing happens with freezing and sublimation.

- Latent Heat of Condensation: It is the amount of heat energy released to the atmosphere when condensation takes place. As mentioned above, when one gram vapour is changed to water 540 calories is released, it is called latent heat of condensation, because it is reaching to the atmosphere due to the process of condensation.

- Latent Heat of Vaporization: It is the amount of heat energy needed to evaporate water and it is taken from the surroundings. When one gram water turns into vapour, 540 calories is utilized. This amount of heat is called latent heat of vaporization.

- Latent Heat of Sublimation: It is the amount of heat required to change solid (ice) directly into vapour without turning into liquid (water) or vice versa. This amount is 620 calories for changing one gram of ice into vapour or one gram of vapour to ice.

- Latent Heat of Freezing/ Melting: It is the amount of heat required to change one gram of water into ice or one gram of ice into water. This amount is 80 calories per gram of water/ ice.

Expansion and Compression of the Air

We have already discussed that the air pressure keeps on declining progressively with increase in height. Consequently, the mass of air is lesser with increase in height. The mass is greater downward. It is due to this reason, any parcel of air, if it rises, going upward is expanded. Because the rising air is entering in the zone of less dense air, it results in expansion. Therefore, the expansion is a natural event in this case. Contrary to this process, when any parcel of air descends, it enters in the zone of more dense air. This results into the compression of the descending air. Rising air expands and the intermolecular space is expanded and it causes the cooling in the air as well. It is also known as adiabatic cooling. That means, the cooling is caused by simply expansion of the volume of air. There is no exchange of heat with the rising air and its surroundings.

Lapse Rate

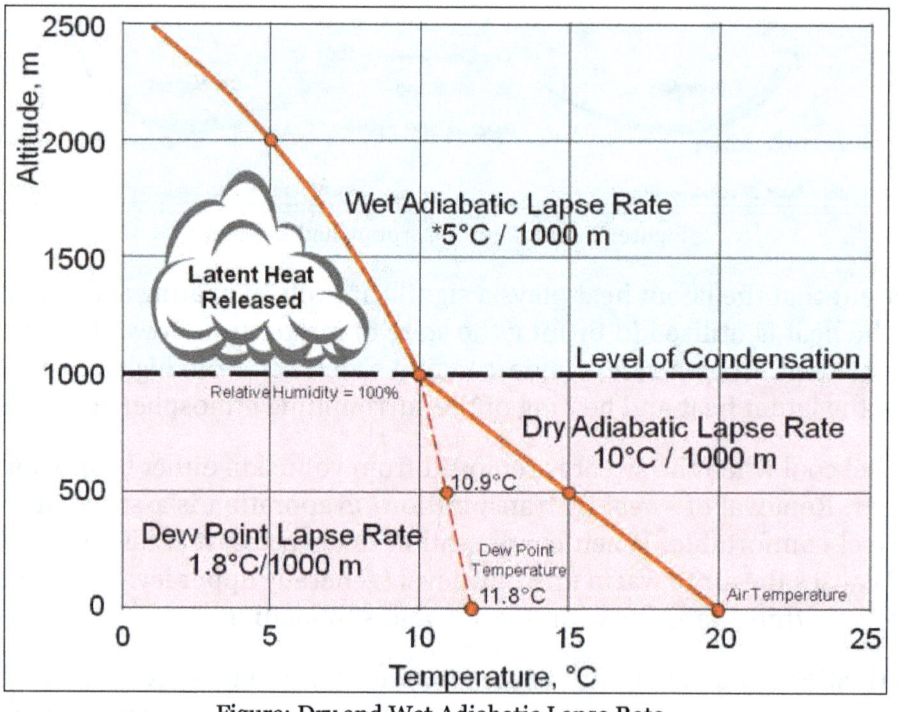

Figure: Dry and Wet Adiabatic Lapse Rate.

Lapse rate is observed change in temperature when moving upward in the troposphere. It is generally counted as drop in temperature with per km ascent in the atmosphere. We know that the mountains are colder than plains if the latitude remains the same. It is due to the lesser effectiveness of the long wave radiation with height.

- Environmental Lapse Rate: It is a simple drop in temperature. We are simply climbing the height and we feel the drop in temperature. It is known as environmental lapse rate. The average environmental lapse rate is 6.5 °C per km. This rate is applicable only in the lower atmosphere (troposphere). It happens due to two reasons, fall in atmospheric pressure and decrease in greenhouse gases (carbon dioxide and water vapour).
- Adiabatic Lapse Rate: In adiabatic lapse rate, the parcel of the air is lifted and the lifting air is forced to expand. Expansion leads to drop in temperature. In this case, air itself is

rising whereas in environmental lapse rate the air was static but someone is going up and feels the declining temperature. Adiabatic lapse rate is categorized into two – dry and wet.

- Dry Adiabatic Lapse Rate: When the airis dry or having very less moisture in it and it rises, the temperature will fall more sharply. Since there is less possibility of condensation, the decline in temperature is at higher side and it is about 10oC per km. In this case the air is almost stable and in general represents high pressure conditions.
- Wet Adiabatic Lapse Rate: When the air is moist, it is unstable. The capacity to hold more moisture in the air is less. There is greater possibility of condensation with rise of air parcel. Once it is fully saturated and the condensation takes place, the latent heat of condensation is released. The released heat is utilized to warm the surrounding air. Therefore, the decrease in temperature is lowered. On an average, the drop in temperature is about 5 °C per km.

Factors Affecting Heating and Cooling of Atmosphere

The earth is a sphere and the atmosphere is encircling around it. The distribution of energy on the earth surface and in atmosphere is varying to a great degree particularly with respect to latitude. The distribution of heat is affected by several factors important among them are:

- Latitude,
- Altitude and nature of earth's surface,
- Differential heating and cooling of land and water,
- Nature of ocean currents,
- Transparency of the sky,
- Slope aspects.

Latitude

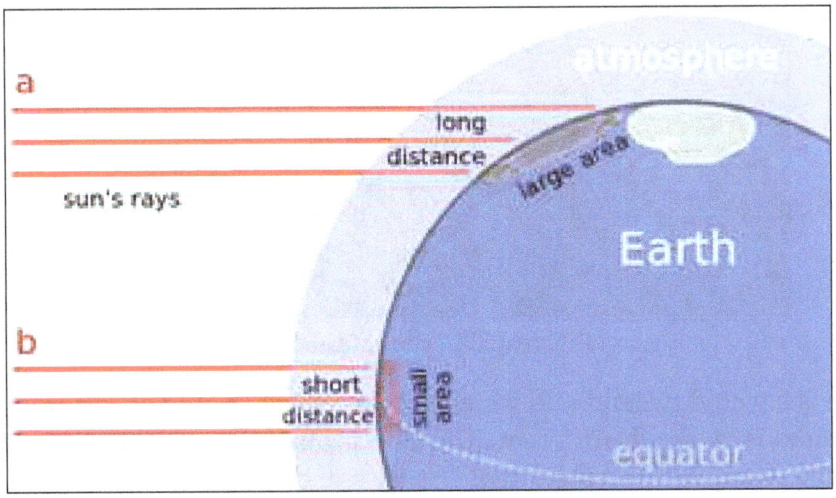

Figure: Decreasing Effectiveness of Solar Energy with Latitudes.

The light energy of the sun reaches to the earth's surface to a maximum limit to 1800 of angle. You know it very well that the earth is a sphere. The angle of incidence keeps on reducing poleward from the equator. Reduction in angle of incidence reduces the energy received per unit of the area. It happens because a beam with vertical or near to vertical will spread to a smaller area while greatly inclined rays will spread over a large area. Therefore, energy received by vertical or near to vertical rays will be greater in comparison to the greatly inclined rays. Sun rays are vertical in equatorial region while in polar region, it is greatly inclined. That is why, the low latitude areas are warmer and high latitude areas are colder. On the other hand, high latitude areas are much colder because of less effective heating.

Altitude and Nature of Earth's Surface

You know it very well that, in general, the mountains are cool and the plains are warm/ hot if the latitude is the same. Plain possess thick layer of atmosphere and the mountain has a small thickness of it. Air is greatly compressed on the plains in comparisons to the mountains. The atmosphere is normally warmed by the long wave terrestrial radiations. Hence, low altitude area has more temperature than the high altitude areas.

The nature of the rock also affects the atmospheric heating and cooling. The areas possessing bare rocks have more intense heating by the sun's energy. That type of area also radiates back more and more received energy and the result is quick heating of the air laying there. On the other hand, the area possessing more and more vegetative cover, the heating is moderate as some of the energy is released from the earth surface in the form of latent heat with evapotranspiration. Vapourazitation form any area keeps it cool. That is why, the forested or grasslands are cooler than the hot deserts. Its effects are also seen on the air/ atmosphere of the area.

Differential Heating and Cooling of Land and Water

Figure: Differential Heating of Land and Water.

The earth surface is covered by land and water bodies. However, the absorption of heat energy to both of these surfaces is markedly different. The land is an opaque to the incoming solar radiation while water is translucent. Whatever the heat energy reaches to the land body, it is utilized to heat a thin layer of the land surface while the same amount of heat energy reaching to the water surface is penetrating to much deeper depths. Apart from that, evaporation also reduces the effectiveness of heat on water surface as a part of its energy is removed by latent heat of evaporation. That is why, we find equalizing climatic condition on the sea coast while the interior part of the continent reflect harsh climate i.e. very cold or very hot. Accordingly the heating and cooling of the air of that region is affected.

Nature of Ocean Currents: Heating and Cooling

Low latitude areas are warmer while the high latitude areas are colder. The temperature of the ocean water is also affected by the temperature distribution over the globe. Ocean currents are flowing under the influence of planetary wind system as well as the regional shape of the sea coasts. Since the ocean currents are very important medium of heat transfer through advection from low latitude to the high latitude. Depending upon the nature and temperature of the ocean water, currents are categorized into two – warm and cold. Wherever those currents are reaching, they are influencing the areas accordingly. That is way; the north-western European coasts are warmer but the north-eastern coast of North America is chilled during winter though both of them lie on the same latitude.

Transparency of the Sky

Apart from the gases, several other minute suspended particles and water vapour constitute the atmosphere. Though the gases are almost uniformly distributed, but other substances are varying at local and regional level. Their availability and quantity is season dependent also. These substances are obstructing, reflecting and absorbing the solar radiations and affecting the heating and cooling of the atmosphere.

Slope Aspects

Slope of the any region or mountain has direct bearing on the heating or cooling of the air. The south facing slope of northern hemisphere and north facing slope of southern hemisphere receive more energy than their counterpart. Therefore, they are warmer and support more rain resulting in dense vegetation cover. When it is the reverse case, it is colder, less moist and retarded growth of vegetation. The examples can be had from the Himalayas.

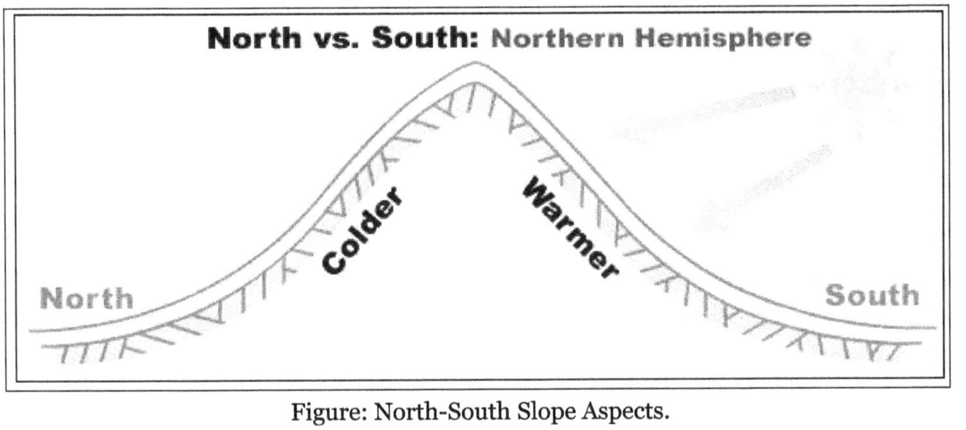

Figure: North-South Slope Aspects.

Eastern and western slope also have bearing on the heating and cooling of the air along them.

Eastern facing slope receive less energy as the temperature is not that effective. Therefore, it is cool on the eastern slope. On the western facing slope, more temperature is recorded and hence, it is warmer. This type of example can be had from the two slopes of the Rockies and the Andes.

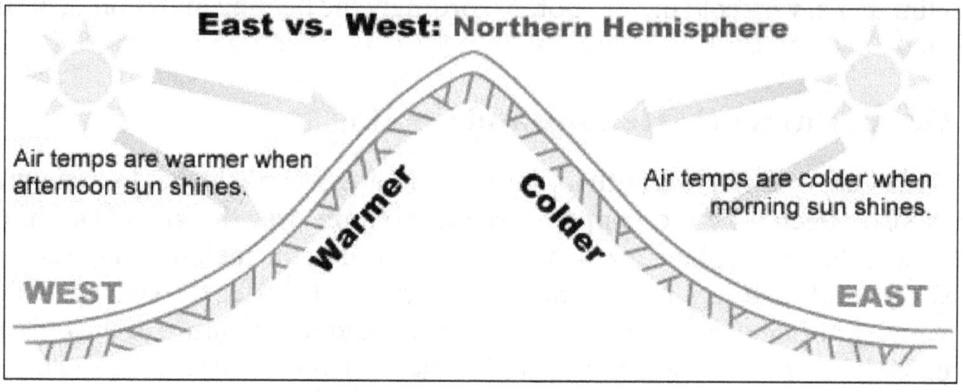

Figure: East-West Slope Aspects.

Insolation

The earth's atmosphere is very much a dynamic entity. Large volumes of air are continually being moved both up and down and across the face of the Earth. As a proof, we feel air when it is in motion. There must be some energy involved here. It needs to be understood that the atmosphere is not a closed system. It is in contact with both the earth and with space, and receives energy from both directions. However, Earth itself directly contributes only a negligible amount of energy to the atmosphere, and its main role is to reflect energy from elsewhere. The ultimate sole source of atmospheric energy is in fact heat and light received through space from the Sun. This energy is known as solar insolation.

The Earth receives only a tiny fraction of the total amount of Sun's radiations. Only two billionths or two units of energy out of 1,00,00,00,000 units of energy radiated by the sun reaches the earth's surface due to its small size and great distance from the Sun. The unit of measurements of this energy is Langley (Ly). On an average the earth receives 1.94 calories per sq. cm per minute (2 Langley) at the top of its atmosphere.

Factors Influencing Insolation

The insolation received on earth is not same everywhere. The amount and the intensity of insolation vary from place to place, during a day, in a season and in a year. The factors that cause these variations in insolation are:

- Revolution of earth around sun: Earth revolves in an elliptical orbit around the sun. Thus, distance between the Sun and the earth vary. The earth is farthest from the sun on 4th July. This position of the earth is called aphelion. On 3rd January, the earth is the nearest to the sun. This position is called perihelion. Therefore, the annual insolation received by the earth at perihelion is slightly more than the amount received at aphelion. However, the effect of this variation in the solar output is masked by other factors like the distribution of land and sea and the atmospheric circulation. Hence, this variation in the insolation does not have great effect on daily weather changes on the surface of the earth.

- The rotation of earth on its axis: Earth rotates around its axis and makes an angle of 66½ with the plane of its orbit round the sun. This particular characteristic of earth has great amount of influence on the amount of insolation received at different latitudes. The seasons in each hemisphere are dictated not by the closeness to the sun but by the axial tilt of the earth.
- The angle of inclination of the sun's rays: Since the earth is round, the sun's rays strike the surface at different angles at different places. The angle formed by the sun's ray with the tangent of the earth's circle at a point is called angle of incidence. It influences the insolation in two ways as follows:
    - When the sun is almost overhead, the rays of the sun are vertical. The angle of incidence is large. Hence, they are concentrated in a smaller area, giving more amount of insolation at that place. If the sun's rays are oblique, angle of incidence is small and sun's rays have to heat up a greater area, resulting in less amount of insolation received there.

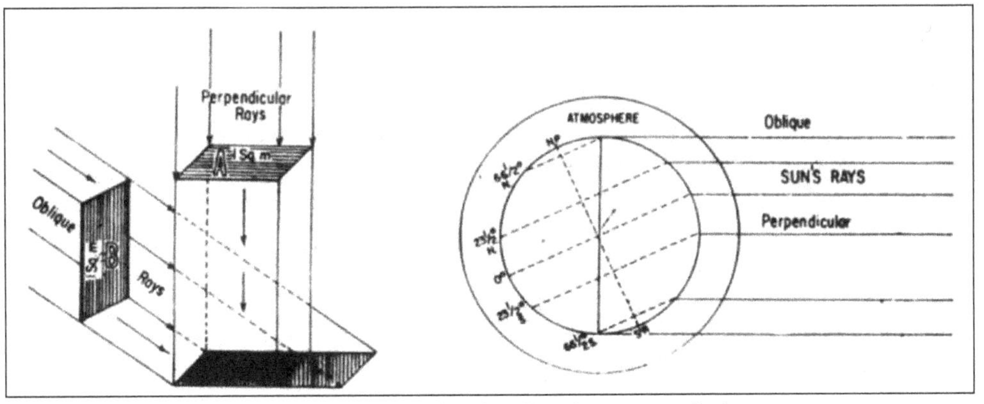

Figure: Effect of angle of inclination on Insolation.

    - The sun's rays with small angle traverse more of the atmosphere than rays striking at a large angle. Longer the path of sun's rays, greater is the amount of reflection and absorption of heat by atmosphere. As a result the intensity of insolation at a place is less. Angle of inclination of solar radiation depends on latitude of a place. The higher the latitude the less is the angle they make with the surface of the earth resulting in slant sun rays. Figure show the winter Solstices in the Northern Hemisphere where angle of inclination is zero at 66 ½ N latitude.

Table: Length of Day on winter and summer Solstices in the Northern Hemisphere.

Latitude	0°	20°	40°	60°	90°
December 22 (winter solstice)	12h 00m	10h 48m	9h 8m	5h 33m	0 m
June 21(summer Solstice)	12h	13h 12m	14h 52m	18h 27m	6 months

- The length of the day: The duration of day is controlled partly by latitude and partly by the season of the year. The amount of insolation has close relationship with the length of the day. It is because insolation is received only during the day. Other conditions remaining the same, the longer the days the greater is the amount of insolation. In summers, the days being longer the amount of insolation received is also more? As against this in winter the days are shorter the insolation received is also less. On account of the inclination of the earth on its axis at an angle of 23 ½ o, rotation and revolution, the duration of the day is not same everywhere on the earth. At the equator there is 12 hours day and night each throughout the year. As one moves towards poles duration of the days keeps on increasing or decreasing. It is why the maximum insolation is received in equatorial areas. Table show the duration of day (in hours & minutes) on winter and summer solstices in the Northern hemisphere.

- The transparency of the atmosphere: The earth's atmosphere is more or less transparent to short wave solar radiation which has to pass through the atmosphere before striking the earth's surface. The transparency depends upon cloud cover, its thickness, water vapour and solid particles, as they reflect, absorb or transmit insolation. High energy ultra-violet rays are absorbed by the Ozone layer. Thick clouds hinder the insolation to reach the earth while clear sky helps it to reach the surface. Water vapour absorbs insolation, resulting in less amount of insolation reaching the surface. Very small-suspended particles in the troposphere scatter visible spectrum both to the space and towards the earth surface.
- Solar variation: It is the change in the amount of radiation emitted by the Sun. These Student Notes: variations have periodic components, the main one being the approximately 11- year sunspot cycle. Sunspots are temporary phenomena on the photosphere of the Sun that appear visibly as dark spots compared to surrounding regions. When there is an increase in sun spots it leads to increase1 in the amount of solar radiation. But this change is almost negligible.

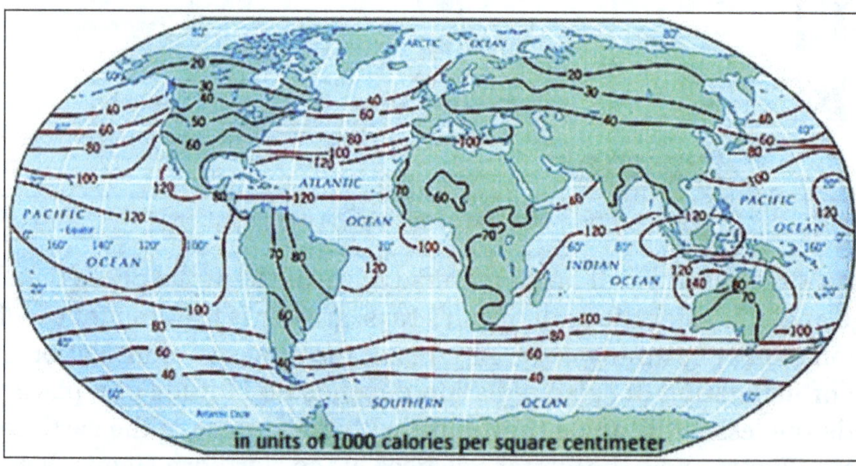

Figure: Average annual insolation on the surface of the earth.

- Topographical variations: Earth does not have a featureless surface. The topographical variations are the major factors modifying the distribution of insolation. Variability in elevation, surface orientation (slope and aspect), and obstruction by surrounding topographic features creates strong local gradients of insolation. Similarly, in the northern hemisphere a south-facing slope (more open to sunlight and warm winds) will therefore generally be warmer and dryer due to higher levels of evapotranspiration than a north-facing slope. This can be seen in the Swiss Alps, where farming is much more extensive on south-facing than on north-facing slopes. In the Himalayas, this effect can be seen to an extreme degree, with south-facing slopes being warm, wet and forested, and north-facing slopes cold, dry but much more heavily glaciated. Vegetation, human activities are more visible on the slopes where insolation is more relatively.

Under combined effect of the above discussed factors, the amount of total annual insolation received by different regions is different. The insolation received at the surface varies from about 320 Watt/m^2 in the tropics to about 70 Watt/m^2 in the poles. Maximum insolation is received over the subtropical deserts. Equator receives comparatively less insolation than the tropics due to presence of clouds. Generally, at the same latitude the insolation is more over the continent than over the oceans because more clouds over the oceans reflect sun rays back into space. Isohels are lines connecting points on the earth surface that receive equal amounts of sunshine. Isohels are more or less parallel to latitudes, especially in southern hemisphere.

Distribution of Insolation

The total incoming solar radiation as known as insolation. It is always high at pole in summer solstices. There are two main factors i.e. sun angle and length of daylight that influenced the insolation in any given location the most. The highest average daily insolation found over the Arctic because of 24 hours daylight. The earth's surface receives different amount of insolation according to season. It receives more insolation in summer than winter because of longer day length in higher altitudes. The upper atmosphere over the north pole is marked maximum of solar radiation at the June solstice, but only 30 per cent of total insolation is absorbed by the surface due to high average cloudiness over the arctic in summer and very high reflectivity (albedo) from the ice surface and snow as compared with the global average of 48 per cent of insolation being absorbed at the surface.

The horizontal distribution of insolation is also varying on earth surface. The maximum temperature is found in the tropics rather than equator because apparent migration of the vertical sun is relatively rapid between 6^0 N to 6^0 S. The vertical sun rays, during autumn and spring equinoxes, lies only for 30 days, allowing very little time span for heating the earth surface very high. Contrary to this, the latitude between 17^0 to 23.5^0 N-S, the insolation occurs almost vertically for 86 consecutive days during solstice. This longer duration of insolation over tropics than at equator, makes the maximum temperature zone at tropics Figure. The distribution of land and water also affect the temperature in this zone. In northern hemisphere, continent enhanced the rise in temperature than oceanic southern hemisphere. Another factor is cloudiness, in sub-tropical high-pressure zone, due to low cloud cover allows large annual receipts of solar radiation.

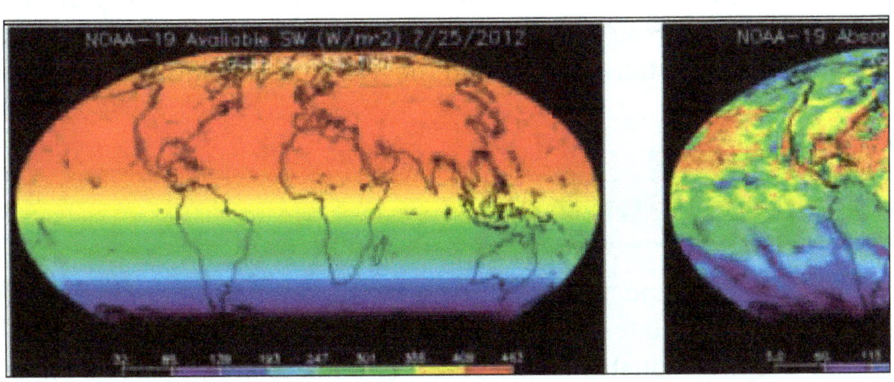

Figure: Available Solar Energy and Energy Absorbed by the Earth.

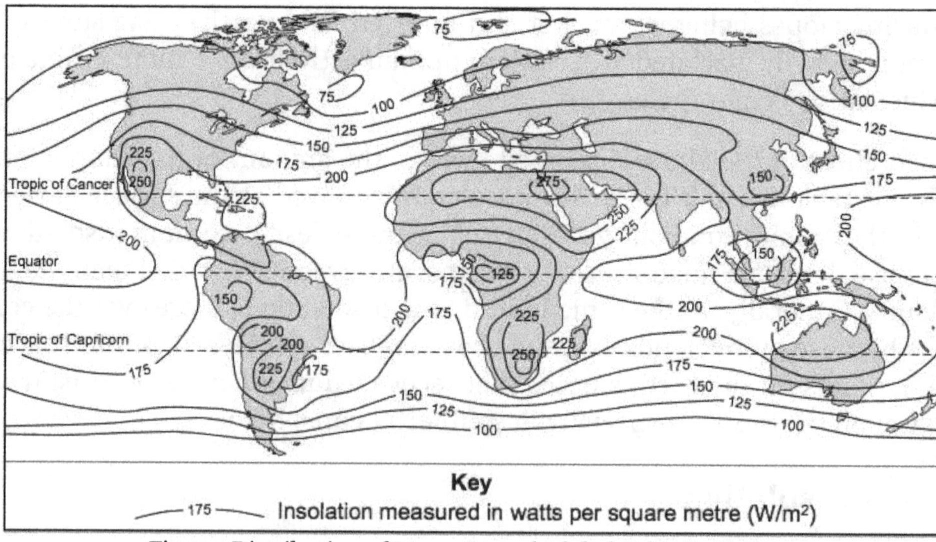

Figure: Distribution of Mean Annual Global Solar Radiation.

Insolation on the earth is received almost entirely from the sun which comes in the form electromagnetic radiation from the surface whose temperature is about 6000 K. The solar constant reaching to the earth is approximately 1368 Wm^2. The amount of energy, emitted by the sun in the form of electromagnetic radiation, received by the earth is very small in comparison to the total energy released from the sun. But it is sufficient to run the earth systems and the biotic life evolved on the earth. The sun's energy is reaching to the earth is basically solar radiation. There are five methods of insolation such as transmission, scattering, refraction, absorption and reflection. These mechanisms impact on the uneven distribution of insolation in the earth. There are several astronomical and geographical factors that also affecting the insolation on the earth. Important among them are solar output, incidence angle of the sun rays, and length of day, variations in distance between earth and sun and transparency of the atmosphere.

The variations of insolation are also due to the solar flares and sun spots that periodically change the patterns. The varying incidence angle of the sun rays and length of day affects the latitudinal variation in the amount of insolation on the earth surface. Consequently, within the tropics, the total insolation is more than higher latitudes. The distance between earth and the sun as well as inclination angle of the earth determine the different season on earth. Therefore, uneven distribution of insolation is very important to understand the atmospheric dynamics and distribution of climate and vegetation on earth surface. The climatic elements such as temperature, precipitation and winds are controlled by the amount of insolation. The livelihood of the people and their comforts as well as plants lives are totally depends on the available insolation on earth surface.

Atmospheric Circulation

The circulation of wind in the atmosphere is driven by the rotation of the earth and the incoming energy from the sun. Wind circulates in each hemisphere in three distinct cells which help transport energy and heat from the equator to the poles. The winds are driven by the energy from the sun at the surface as warm air rises and colder air sinks.

The circulation cell closest to the equator is called the Hadley cell. Winds are light at the equator because of the weak horizontal pressure gradients located there. The warm surface conditions result in locally low pressure. The warm air rises at the equator producing clouds and causing instability in the atmosphere. This instability causes thunderstorms to develop and release large amounts of latent heat. Latent heat is just energy released by the storms due to changes from water vapor to liquid water droplets as the vapor condenses in the clouds, causing the surrounding air to become more warm and moist, which essentially provides the energy to drive the Hadley cell.

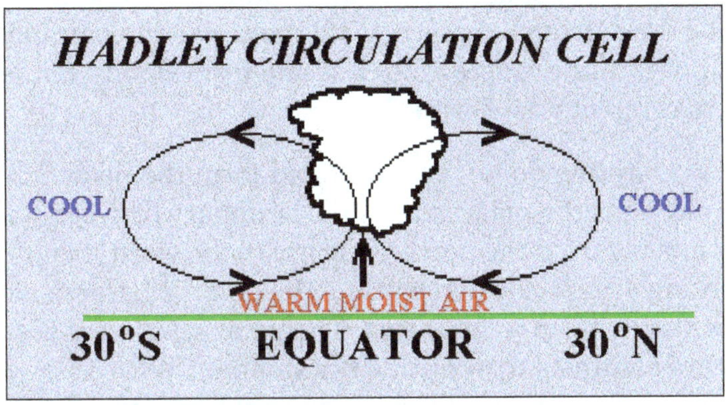

Figure: Hadley Cell circulation.

The Hadley Cell encompasses latitudes from the equator to about 30°. At this latitude surface high pressure causes the air near the ground to diverge. This forces air to come down from aloft to "fill in" for the air that is diverging away from the surface high pressure. The air flowing northward from the equator high up in the atmosphere is warm and moist compared to the air nearer the poles. This causes a strong temperature gradient between the two different air masses and a jet stream results. At the 30° latitudes, this jet is known as the subtropical jet stream which flows from west to east in both the Northern and Southern Hemispheres. Clear skies generally prevail throughout the surface high pressure, which is where many of the deserts are located in the world.

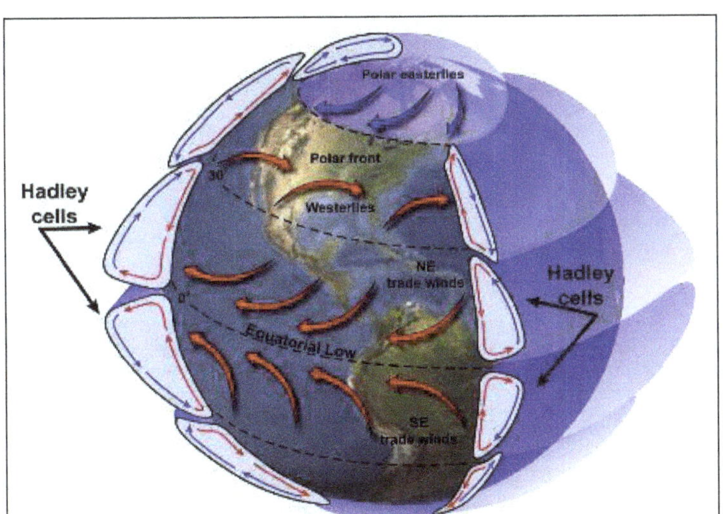

Figure: General Wind Directions.

From 30° latitude, some of the air that sinks to the surface returns to the equator to complete the Hadley Cell. This produces the northeast trade winds in the Northern Hemisphere and the southeast trades in the Southern Hemisphere. The Coriolis force impacts the direction of the wind flow. In the Northern Hemisphere, the Coriolis force turns the winds to the right. In the Southern Hemisphere, the Coriolis force turns the winds to the left.

From 30° latitude to 60° latitude, a new cell takes over known as the Ferrel Cell. This cell produces prevailing westerly winds at the surface within these latitudes. This is because some of the air sinking at 30° latitude continues traveling northward toward the poles and the Coriolis force bends it to the right (in the Northern Hemisphere). This air is still warm and at roughly 60° latitude approaches cold air moving down from the poles. With the converging air masses at the surface, the low surface pressure at 60° latitude causes air to rise and form clouds. Some of the rising warm air returns to 30° latitude to complete the Ferrel Cell.

The two air masses at 60° latitude do not mix well and form the polar front which separates the warm air from the cold air. Thus the polar front is the boundary between warm tropical air masses and the colder polar air moving from the north. (The use of the word "front" is from military terminology; it is where opposing armies clash in battle.) The polar jet stream aloft is located above the polar front and flows generally from west to east. The polar jet is strongest in the winter because of the greater temperature contrasts than during the summer. Waves along this front can pull the boundary north or south, resulting in local warm and cold fronts which affect the weather at particular locations.

Above 60° latitude, the polar cell circulates cold, polar air equatorward. The air from the poles rises at 60° latitude where the polar cell and Ferrel cell meet, and some of this air returns to the poles completing the polar cell. Because the wind flows from high to low pressure and taking into account the effects of the Coriolis force, the winds above 60° latitude are prevailing easterlies.

Walker Circulation

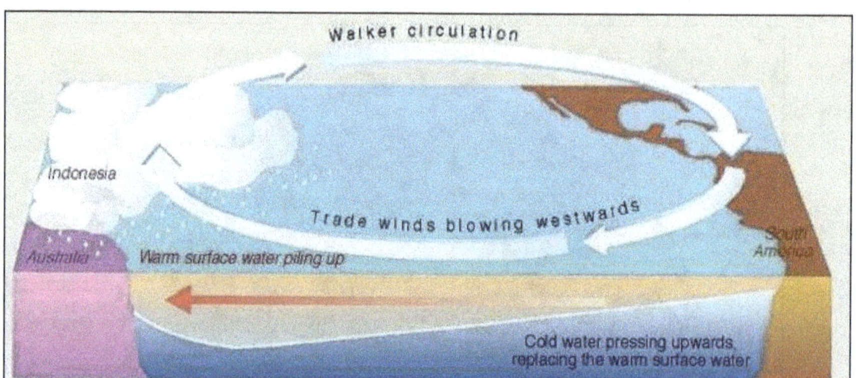

Figure: Walker Circulation.

In contrast to the Hadley, Ferrel and polar circulations that run along north-south lines, the Walker circulation is an east-west circulation. Over the eastern Pacific Ocean, surface high pressure off the west coast of South America enhances the strength of the easterly trade winds found near the equator. The winds blow away from the high pressure toward lower pressure near Indonesia. Upwelling, the rising of colder water from the deep ocean to the surface, occurs in the eastern Pacific along South America near Ecuador and Peru. This cold water is especially nutrient-rich and is stocked with an abundance of large fish populations. By contrast the water in the western Pacific, near Indonesia, is relatively warm. The air over Indonesia rises because of the surface low pressure located there and forms clouds. This causes heavy precipitation to fall over the western tropical Pacific throughout the year. The air then circulates back aloft towards the region above the surface high pressure near Ecuador and this becomes the Walker circulation. The air sinks at this surface high pressure and is picked up by the strong trade winds to continue the cycle.

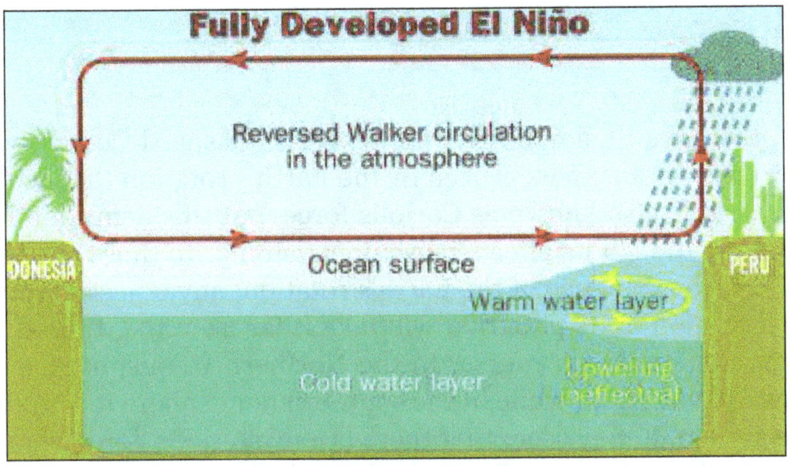

Figure: El Nino.

On some occasions, the Walker circulation and the trade winds weaken, allowing warmer water to "slosh back" towards the eastern tropical Pacific near South America. You can think of this as blowing a fan over a bathtub full of water. If the fan blows steadily, the water at the side farthest from the fan will tend to pile up downwind. If you suddenly slow the fan down, some of the water that was built up will surge back towards the fan. The warmer water will cover the areas of upwelling, cutting off the flow of nutrients to the fish and animals that live in the eastern Pacific Ocean. This warming of the eastern Pacific Ocean is known as El Niño. The warmer water will also serve as a source for warm, moist air which can aid in the development of heavy thunderstorms over the mass of warm water.

Moving Masses of Air

It is difficult to construct a model of atmospheric circulation, because very many factors influence the displacement of masses of air through the atmosphere; however, the fundamental principle is that the gases that make up the atmosphere tend to seek a condition of equilibrium that implies a uniform distribution of energy, equalizing the temperature and the pressure on the entire planet. The 'motor' of atmospheric circulation is therefore given by the redistribution of the energy received from the Sun. Solar radiation, in fact, varies at different latitudes with the result that the equatorial regions are warmer than the polar ones. The atmosphere tends to re-balance this difference moving masses of warm air from the regions where there is an excess of heat towards colder regions, in an attempt to reduce the difference in temperature between the Equator and the Poles.

Differences in temperature bring about immediate differences in the pressure within the masses of air: it is these differences in pressure that cause air displacement. Low pressure areas attract air from areas where the pressure is higher. Vice versa, air tends to move away from high pressure areas towards areas with lower pressure. The speed at which this displacement takes place is directly proportional to the difference in pressure between the two points. In ideal conditions, if the Earth were motionless and if there were no friction or obstacles, air would flow perpendicularly to the isobars (the lines that join points with equal pressure), according to the so-called pressure gradient, following a path that makes the air travel along the shortest route from the high pressure area to the low pressure one. However, many factors contribute to deviating the flow of air from the ideal condition.

The Coriolis Effect

The Coriolis Effect, that gets its name from the physicist Gaspard Coriolis who postulated the theory in the 19th century, is an effect, caused by the Earth's rotation that acts on moving bodies on the Earth's surface. It is also known as Coriolis force, but this definition is incorrect since it is not really a force, even though its effects may appear similar to those of a force to an observer within the system. As a consequence of the Earth's rotation, any object that travels in a straight line will undergo a deviation in its path that will take it far from the desired point unless it can carry out continuous changes in its course. In the Northern Hemisphere, every moving object will tend to drift towards the right, while in the southern hemisphere it is towards the left. To an observer within the system, it may seem that there is a mysterious 'force' that pushes objects off their course, but in actual fact, it is an effect caused by 'displacement' of the Earth's surface under the moving object.

The Coriolis Effect is proportional to the angular velocity of the Earth (that is constant), to the speed of the moving object and to the latitude: it is null at the Equator and maximum at the Poles. This effect can be observed in many daily situations: in the Northern Hemisphere aeroplanes have to change their course towards the left, the right railway track undergoes a greater wear, etc. Similarly, even winds undergo deviations: if the Earth were motionless, the winds would tend to move in an N-S direction, from the Equator to the Poles along the meridians, but as a result of the Coriolis Effect, winds from the North tend to become winds from the NE and so on.

High and Low Pressure

If an isobar chart is observed, it can be seen that pressure is not distributed uniformly in the atmosphere around our planet: there are areas with a lower pressure than the surrounding areas and areas where the pressure is higher. Due to a characteristic of gases, air tends to move from high pressure areas towards those with low pressure in an attempt to balance the difference. The presence of high and low pressure areas is therefore the principal motor of all meteorological phenomena, in other words, of the 'weather'. Hence, it is important to understand how air circulates close to these areas and how they are distributed in the atmosphere.

Anticyclones

In high pressure zones, air tends to sink towards the ground causing the air that is present to move away with a divergent movement. The air gets compressed while descending and tends to disperse the clouds, and in fact high pressure conditions are associated with settled and calm weather. As a result of the Coriolis Effect, air tends to move away from the high pressure system, clockwise in our hemisphere and anticlockwise in the Southern Hemisphere (anticyclonic circulation).

Cyclones

A low pressure area, instead, tends to attract air from the surrounding region where the pressure is higher. Near the centre of the cyclone, air tends to rise higher attracting a growing amount of air from the neighbouring areas. On rising, air expands and cools with the subsequent formation of clouds and precipitation: it is for this reason that low pressure areas are usually associated with bad weather. Air tends to converge towards the low pressure centre with an anticlockwise movement in our hemisphere and a clockwise movement in the Southern Hemisphere (cyclonic circulation).

Circulation Cells

Temperature and pressure differences are not distributed casually in the atmosphere but permanent and stable low and high pressure areas can be identified, which are organized so as to form big circulation cells around the world. This situation, obviously, is not static and unchangeable. During the year the circulation cells move towards the North or the South, depending on the unequal amount of solar energy that the different regions of the Earth receive in each season: in our hemisphere, they move towards the Equator in winter and towards the Poles in summer. Three main circulation cells can be identified in each hemisphere that are placed symmetrically respect to the Equator.

When Air Masses Meet

When a moving cold air mass meets a warm air mass, that is lighter, it tends to wedge below the latter, thus giving origin to a cold front. The warm air is forced upwards and its ascent causes the formation of clouds. Since the surface of contact between the two masses is quite steep and ascent is rapid, the clouds will be prevalently of the cumuliform type. The passage of a cold front is accompanied by widespread cloud systems and precipitations, with a sharp drop in the temperature, an increase in the pressure and precipitations often characterized by thunderstorms. On the weather charts, cold fronts are indicated using a black line with triangles. Cold fronts form typically at our latitudes, when the cold dry air from the Polar Regions meets the humid air coming from the tropical regions.

Warm Fronts

If, on the contrary the warm air mass is moving towards the cold air mass, the warm air mass slowly slips over the cold one, and rises along a vast gently sloping surface: in this case we can speak of a warm front which also brings clouds and perturbation systems. The margins of a warm front are less marked that those of a cold front, changes are more gradual and the perturbations arrive less rapidly. The passage of a warm front is marked by a rise in the temperatures, a decrease in pressure and persistent rains, which are however of a moderate intensity. Since the clouds rise slowly, the cloud system that forms generally consists of stratified clouds. On the weather charts, warm fronts are indicated with a black line with semicircles.

Occluded Fronts

An occluded front forms when a cold front reaches a warm front, forcing all the warm air to rise to higher altitudes and the cold air is stratified near the ground. At our latitudes, over the Atlantic, cyclonic areas form continuously; these are fed by the Azores anticyclone and the Polar anticyclone. Here, the tropical warm and humid air and the cold dry polar air meets, and generally a depression vortex is formed in which a warm front and a cold front are active. The cold front is generally more active and advances more rapidly than the warm front. When the cold front reaches the warm front an occluded front is originated. After having discharged the remaining humidity on to the occluded front, the weather generally returns to good weather, the air on the ground starts to warm again: and the perturbation is exhausted. On the weather charts occluded fronts are indicated with a black line with alternating triangles and semicircles.

Perturbation Front

The presence of mountain ranges on the course of a front can provoke variations and deformations of various types that can lead to evolutions in a perturbation that are difficult to forecast.

Circulation Cells

Between the Equator up to 30° latitude (N or S), we find the Hadley cell. In the Equatorial region, air is heated and rises, creating a low pressure area. Air would tend to shift towards the N along the meridians, but in the northern hemisphere, due to the Coriolis effect, the flow is deviated towards NE and descends toward the 30° parallel, bringing warm humid air. Once it descends, the air is again drawn towards the Equator due to the low pressure of the area, and this time the air travels from NE to SW, again as a result of the Coriolis Effect. This movement gives origin to the NE trade winds. The same occurs in the southern hemisphere where the trade winds blow from the SE. The area where the NE trade winds clash and converge with the SE trade winds creates an equatorial low pressure area characterized by precipitations and violent perturbations, the so-called area of equatorial calm, that gets its name due to the low pressures and low wind speeds. Between the 30° and 60° latitudes, instead, in both hemispheres the Ferrel cell is active. It rotates in the opposite direction to the Hadley cell. Converging with the margin of Hadley's Cell, a tropical high pressure area is created, around 30°, the so-called horse latitudes area, where the winds blow from the SW on ground, the westerlies which however have a more irregular trend than the trade winds. In this belt there are a series of anti-cyclonic nuclei, among which the Azores anticyclone whose seasonal movements determine the weather in our regions. The masses of air from the Ferrel cell move back to the higher altitudes around the 60° latitude, where the area of sub-Polar low pressure is formed.

In the higher latitudes the Polar cell forms, which has the same trend as the Hadley cell with easterlies at ground level and westerlies at higher altitudes? The Polar cells are the least extended, but thanks to the Polar high pressures; these have the important task of transferring the freezing polar air to the middle latitudes in the Ferrel cell.

Factors Affecting Atmospheric Circulation

Pressure Gradient Force Pressure gradient is the change in the air pressure per unit of distance travelled along a certain line. In another words, it is an average change in barometric pressure per unit of distance along a certain linear direction in a given region. From this figure, it is obvious that the pressure gradient determines the direction as well as the intensity of the wind blow. The winds always blow from the high pressure areas to the low pressure areas. Apart from the direction, its velocity is the product of the pressure gradient.

Pressure gradient can be calculated by the change in the isobaric values (barometric pressure generally in mili bars per km (distance)). For example, suppose the two places X and Y are 100 km away from one another. At X, suppose, the air pressure is 1020 mb and at Y, it is 1010 mb. Therefore, the pressure gradient can be calculated by dividing 10 (difference between 1020 mb and 1010 mb) by 100 km. It gives a value of 0.1 mb/km. Take another example, when at A, the air pressure is 1020 mb and at Y, it is 980 mb. Get the difference between 1020 mb and 980 mb. It comes to 40 mb. Divide it by 100 km. The result would be 0.4 mb/km. Out of these two examples, the second one is four times greater than the first one; hence, the wind speed would be four times more in comparison to the first.

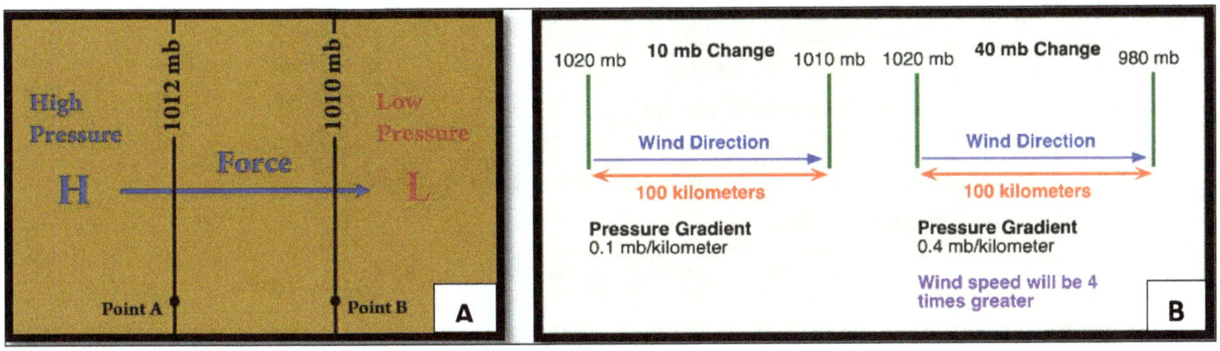

Figure: Air Pressure Gradient and Wind Speed.

Understand that temperature and moisture are the prime factors associated with atmospheric pressure. Temperature distribution depends on insolation and various factors controlling its distribution. Therefore, insolation, temperature and pressure are interlinked.

Coriolis Force

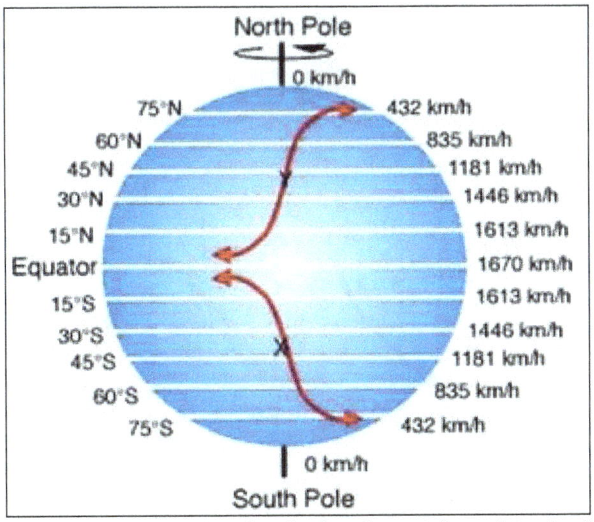

Figure: Earth's Rotational Velocity along Selected Latitudes.

Coriolis, practically speaking is not a force but it an effect which is observed on a mass of body in a rotating system. It results from the rotational movement of the earth and the movement of air in relation to the earth. It acts perpendicular to the axis of the earth. It is determined by the mass of the body and its rate of rotation. The earth rotates from west to east on its axis. Hence, the Coriolis force operates in north-south direction. The Coriolis force is zero at the equator and maximum at the poles. This concept was first explained by French engineer G. G. Coriolis in 1835, thus it is known by his name.

The earth is rotating on its axis. Equator has the maximum bulge. One rotation takes about 24 hours. The velocity of the rotation of the earth is 1670 km/hour (the circumference is about 40000 km) along the equator which comes down to half (835 km/hour) along 60° north and south latitudes. Since the pole is a point and it does not have any circumference, the rotational velocity at pole is zero.

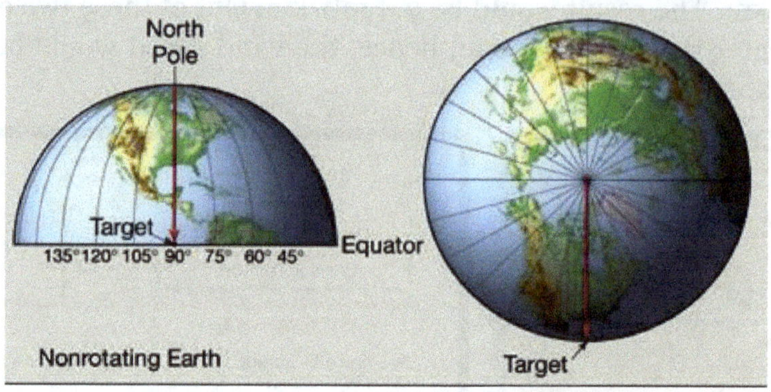

Figure: Movement of Air on Non-Rotating Earth.

When an air moves equator ward in the northern hemisphere on a non-rotating earth it reaches to the destination along the designated path. We know that the earth is not a nonrotating body but, it rotates on its axis. When the air moves equator ward on the spinning earth in the northern hemisphere, the air is deflected to the right as the surface itself moves forward. The same thing also happens when the air is moving from equator to pole, i.e., rightward turning in the northern hemisphere. When you take the case of southern hemisphere, leftward turning of the moving air is observed. The turn is explained by the relative motion/ velocity of the earth along different latitudes. The Figure explains the Coriolis Effect very clearly. Its effect is more visible on the moving objects like the winds, ocean currents or aero planes/ fighter planes.

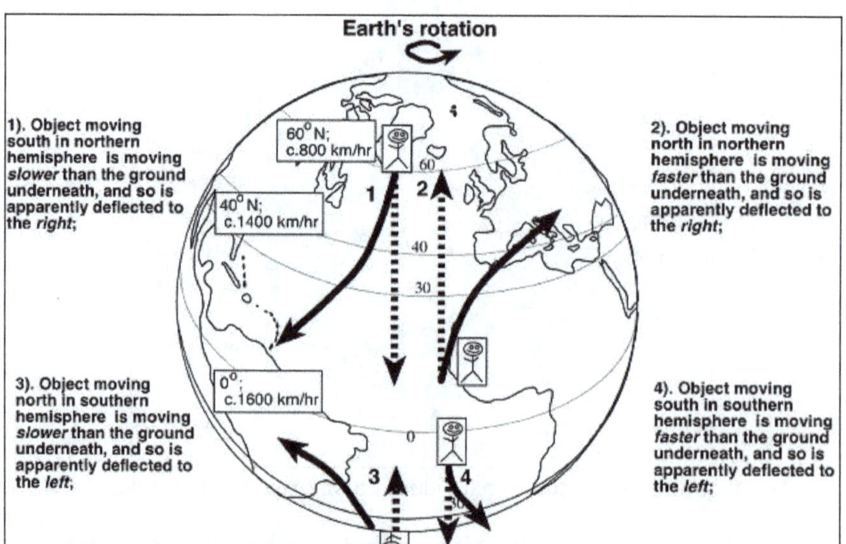

Figure: Movement of Air on Rotating Earth.

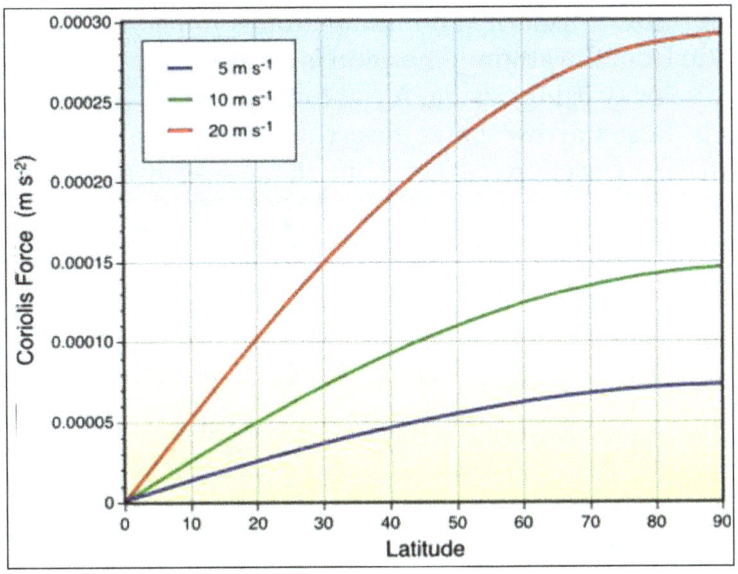

Figure: Wind Velocity and Coriolis Effect.

The Coriolis Effect does not affect the wind speed, but affects the direction. But the wind speed definitely affects the Coriolis Effect. The Coriolis Effect is stronger when the moving air masses is huge and have greater velocity. Therefore, the greater velocity wind is deflected more in comparison to the less strong. All the three lines representing different velocities of the wind are merged together at equator. The violet line shows the wind speed of 5 meter/second (18 km/hour) while the red line shows 20 meter/second (72 km/hour). It is obvious that the differences among these lines are great at the poles. The highest velocity is associated with highest Coriolis force.

Frictional Force

Air friction is the resistance to motion of air in relation to the surface roughness and irregularity through which the wind is blowing. Frictional force reduces the velocity of the wind near the ground. Therefore, wind is most affected by the friction and it is maximum near the ground. As long as we keep on going above the ground, the friction is reduced very drastically.

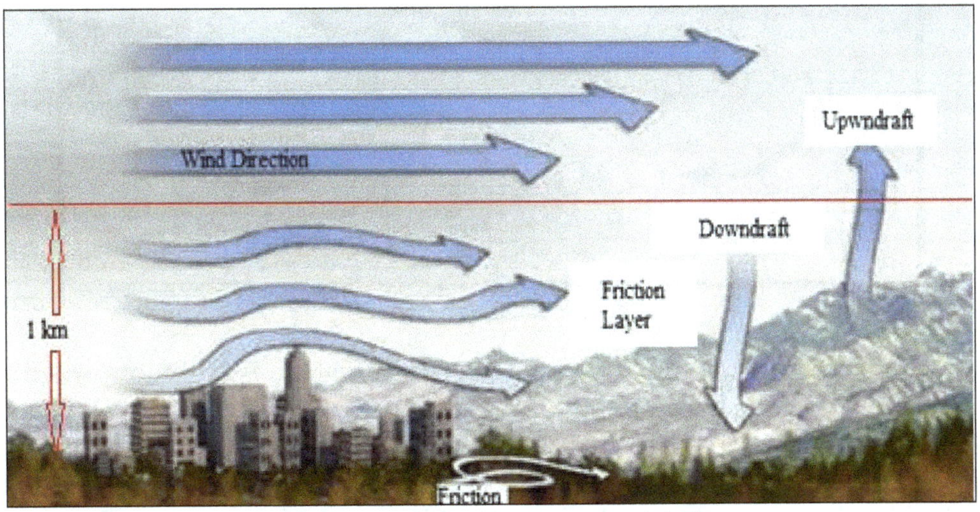

Figure: Friction and its Impact on Winds.

It is obvious that the surface irregularity and undulation is impacting the blowing winds. Within a height of about one km from the ground, the wind is creating eddies and straight blow is heavily affected. After that the velocity is increasing. A right moving arrow shows the direction as well as the velocity of the winds. Bigger arrow shows the greater speed, the smaller one indicate lesser velocity. On the land surface, the friction is greater but on the sea surface due to uniformity, it is less.

Centripetal Force

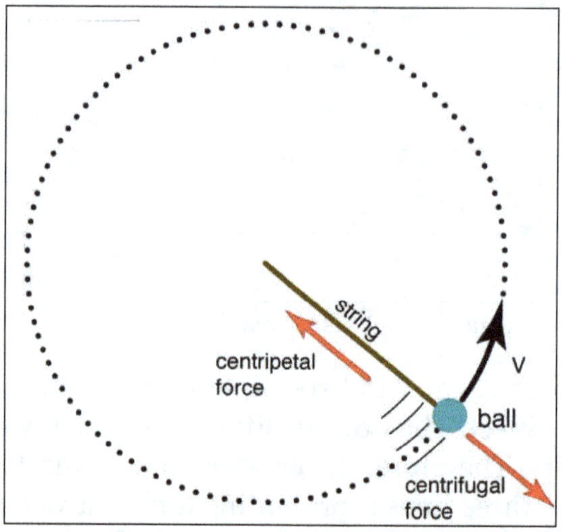

Figure: Diagrammatic Presentation of Centripetal and Centrifugal Forces.

Centripetal force operates at right angles to the blowing wind. It is the inward pulling force i.e. towards the centers of rotation. You must be very well aware that what the centripetal force is. Suppose you fix a ball with a string and you rotate by holding the other side of the string. The resultant forces are centripetal and centrifugal.

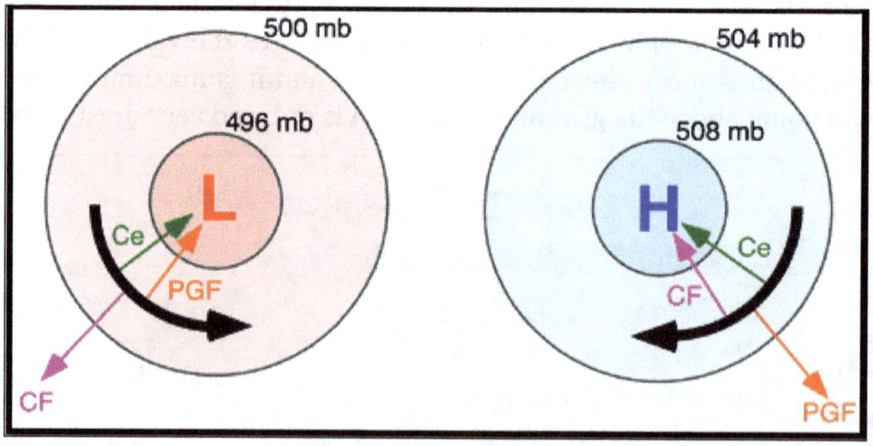

Figure: Balancing Forces.

Generally, the winds are blowing in curved paths. The curvature is still more when the isobaric circular patterns are developed around high pressure or low pressure (anticyclone and cyclone). It is very well developed above the friction layer. Vertically above about a height of around 5.5 km, the air pressure is approximately 500 mb. By this height, the friction force is almost terminated except over the highlands and mountains. Refer to the Figure, which shows that the winds above the friction layer are blowing in a circular manner. Since the friction is over, in case of low pressure, Coriolis force is equal to centripetal force plus pressure gradient force. But in case of high pressure where anticyclone is developed in upper troposphere, pressure gradient force is equal to Coriolis force plus centripetal force. In these conditions, the winds are blowing parallel to the isobars. It is called geostrophic winds by which the jet stream is developed.

Scales of Atmospheric Circulation

The atmospheric circulation happens at a varied scale of operation starting from a very small local scale to the global one. The circulation is classified on the basis of spatial extension as well as its temporal duration. The classification of the atmospheric circulation has been presented in Table.

Category of Scale	Duration	Spatial Extension	Examples
Micro scale	Minutes and seconds	Meters to km	Turbulence, small cumulus Clouds, eddies.
Meso scale	Minutes to hour and to one day	Km to 100s of km	Thunderstorms sea breezes, mountain circulation.
Synoptic scale	Days to weeks	1000s of km	Fronts, cyclones, anticyclones.
Planetary scale	Weeks to months	Global	Planetary winds or permanent winds.

The planetary scales atmospheric as well as ocean circulation are redistributing the thermal energy on the earth surface and atmosphere. The atmospheric circulation varies from one year to another, but the large scale patterns remain almost alike. The synoptic scale movements like frontal development, cyclones etc. have certain duration of occurrence but not very definitely pinpointed one. Their variations are sometimes randomly associated with certain conditional fulfillment. That variation might be of hundreds of km. They develop in specific belts in both the hemispheres.

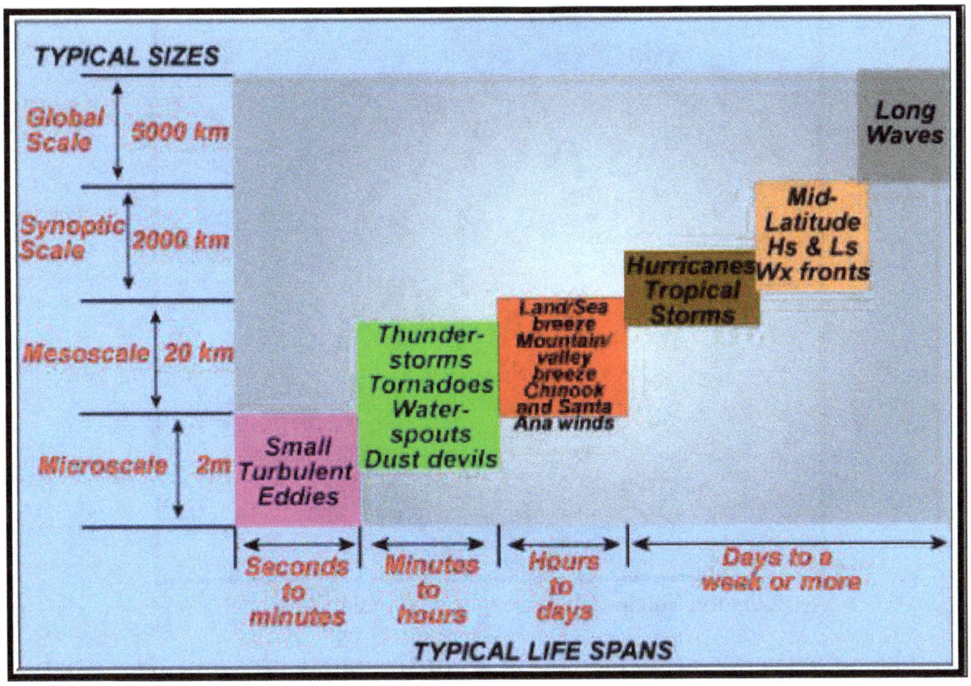

Figure: Spatio-Temporal Atmospheric Circulation.

The meso scale weather events and movements are associated with further small areas and their occurrences are of less than a day. Greater degree of variation is associated with these events. Some of them are directly associated with the heat received from the sun and are most commonly found at a regular diurnal interval. Some are not very predictive and appear randomly, but require fulfillment of certain conditions. The micro scale events are still at further minute level and they are mostly the product of the local variations.

Planetary Atmospheric Circulation

At the planetary scale, the atmospheric circulation is divided into two groups. They are:

- Single-Cell Atmospheric Circulation Model.
- Three-Cell Atmospheric Circulation Model.

Single-Cell Atmospheric Circulation Model

The term "cell" refers to the cycling of the air. Here, we are calling it a single-cell; it means that it is operating at a global scale, but divided at the equator. Hence, in both the hemisphere one cell is found, its occurrence is now debatable. The single-cell atmospheric circulation model was proposed by George Hadley in 1735. At that time scanty information was available about the atmospheric science. Based on his personal experience about more heat in the equatorial belt and very cold polar region, he proposed that the air in the equatorial region is expanded due to more temperature. In expanded air, the air molecules are widely spaced. Therefore, the density is reduced. Reduction in density causes the air to be lighter. Lighter air gets uplifted and moves aloft with upper limit at tropopause.

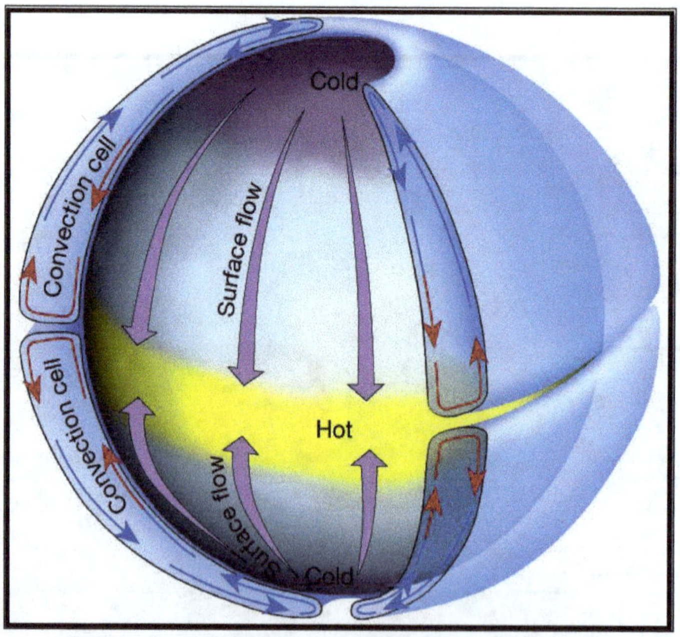

Figure: Single-Cell Atmospheric Circulation Model.

By that height, it becomes very cool. Since, there is continuously uprising air is existing, the uplifted air is being pushed towards north and south in both the hemispheres. Both the poles are very cold; hence, the air over the poles is very dense and heavier. This leads to create high pressure belts surrounding the poles. According to this concept, cold air sinks at the poles and warm and heated air rises at equator. It results into the formation of Hadley cell which is named after its inventor.

Is Single-Cell Model in Operation?

George Hadley proposed the single-cell model, but it would develop only when the sun is overhead at equator and the earth is static, which is not the reality. While proposing this model, he did not consider other factors like apparent shift of the sun's position in an annual cycle. Earth's rotation and Coriolis Effect were also not taken into account by Hadley. The distribution of land and water also play a very significant role in the heat transfer, heating of the surface and the air in their contact. Therefore, the idea of single-cell was well accepted long time back but now the researches have proved that the single-cell in not in operation rather it is three-cell atmospheric circulation.

Three-Cell Atmospheric Circulation Model

The distribution of pressure belts and winds is fairly well established and their seasonal variations are also well known. The pressure belts are result of interplay of various factors. The pressure variations give genesis to horizontal motion of air i.e. winds.

It is well known fact that the maximum sun energy is received in the equatorial region and this energy is redistributed by the winds and ocean currents as mentioned earlier also. But this energy redistribution is not directly from the high energy zone to the low energy zone in a single cell. This redistribution is performed through well-developed distinct three cells. Three-cell formation and their relative regulation is in a sequential manner. These cells divide the troposphere into three separate, distinct and almost closed atmospheric circulations. The three cells are:

- Tropical cell,
- Mid-latitude cell,
- Polar cell.

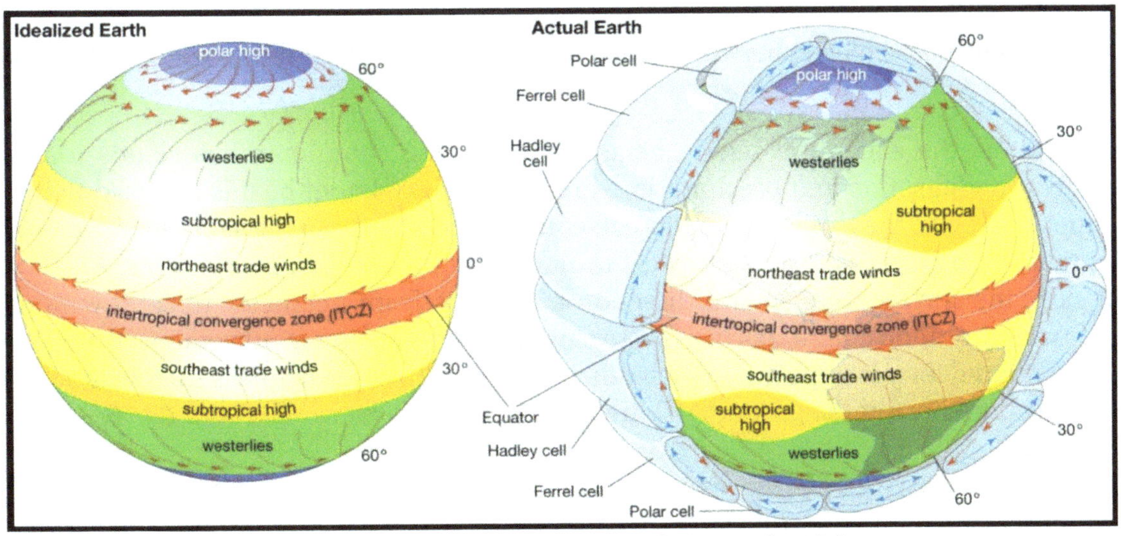

Figure: Three-Cell Atmospheric Circulation on the Globe.

Tropical Cell Tropical cell exists between equator and 300 north and south latitudes. It is thermally induced cell. The upper limit of the cell is demarcated by the tropopause. First of all, it was explained by George Hadleyin 1735 and so it is named after him.

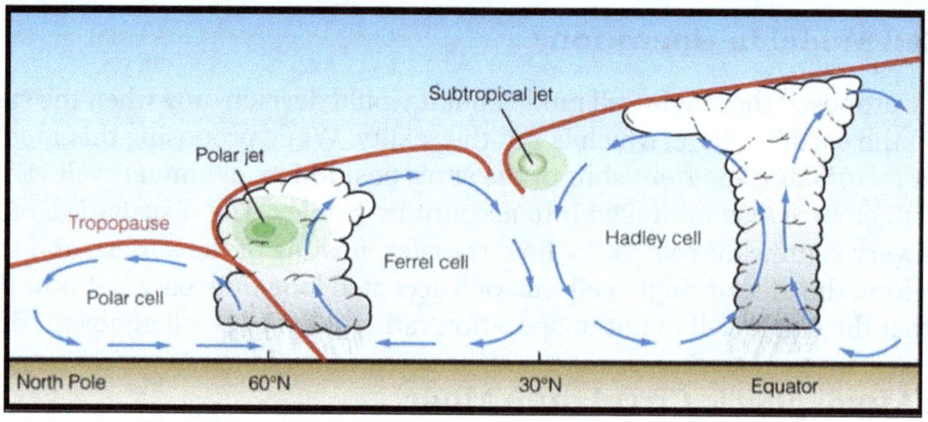

Figure: Cross Sectional Three-Cell Atmospheric Circulation.

Mechanism

The maximum heat is received in equatorial region. Due to high heat, the air expands and it becomes light and rises up. This ascending air cools gradually and starts descending down over around 30 north and south latitudes due to rotation of earth. This descending cool and denser air increases atmospheric pressure in these belts which are also known as horse latitudes. The winds blow from these subtropical high pressure belts to equatorial low pressure belt as trade winds and complete circulation develops in the tropics of both the hemispheres. It is termed as Hadley cell. It has three distinct characteristics:

- The Hadley cell exists throughout the year and is always in operation with slight north and south shifting depending upon seasons.
- The upper air movement poleward conserves its axial angular momentum, but the equatorward moving air faces surface friction and finally loses velocity.
- The balance is achieved between warm upward moving air and the wind converging at the ITCZ.

The rising air in the equatorial region helps in developing cumulonimbus clouds to extensive height. After condensation and precipitation, great amount of latent heat is released there, and it gives impetus to drive the tropical/ Hadley cell equator ward. The direction of the movement of air in the upper troposphere is from southwest to northeast in northern hemisphere. At the ground level or near the surface, the general direction of wind is from northeast to southwest as trade winds of northern hemisphere, baring the seasonal changes. The air movement at upper troposphere and ground level is opposite to each other. In general, the principle is that whatever is wind blowing direction there at the ground, its direction aloft is reverse.

Inter-Tropical Convergence Zone

The trade winds' coming from the sub-tropical highs from both hemispheres converges near equator during equinox. It is this zone from where the air rises and it is generally termed as doldrums. It is a zone of almost no horizontal air movement. With the changing seasons, the departure in the Hadley cell and ITCZ is also observed.

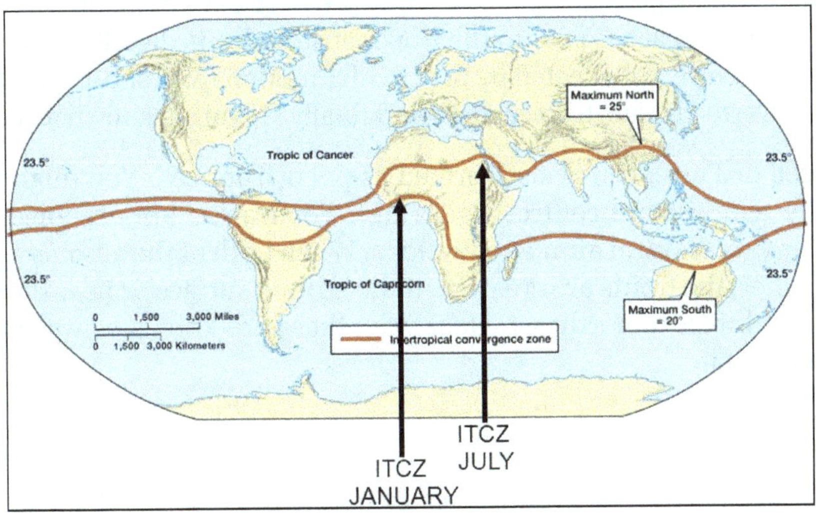

Figure: Seasonal Shift of ITCZ.

Mid-Latitude Cell

Mid-Latitude cell exists between 30 to 60 north and south latitudes. It is found between Hadley cell and polar cell. It is developed due to dynamic conditions prevailing between the two cells bordering it. This cell is operating just reverse to tropical cell. First of all, it was explained by William Ferrel in 1856 and so it is named after him.

Mechanism

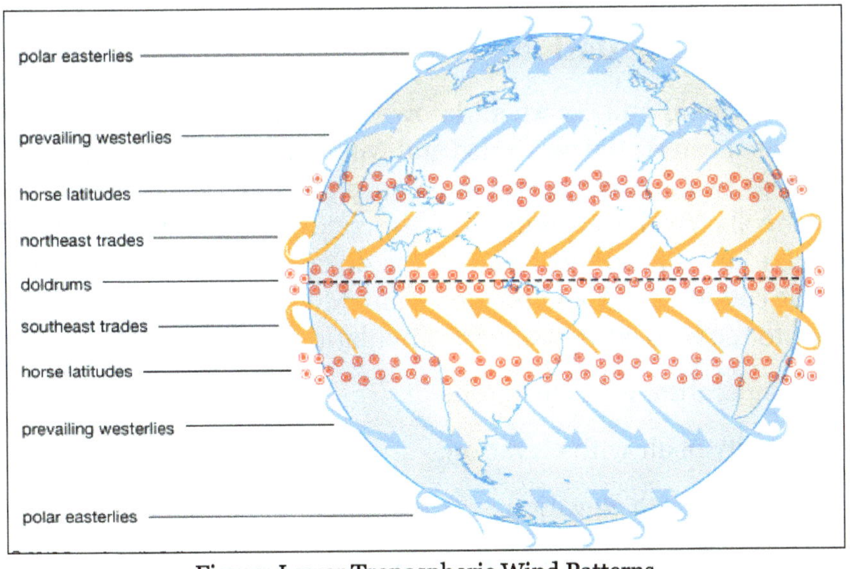

Figure: Lower Tropospheric Wind Patterns.

The low pressure around equator is thermally induced. The high pressure at subtropical zone is dynamically created by subsiding air from above coming from the ITCZ. From this high pressure, winds are blowing in both the directions – towards equator and towards poles. But before reaching the pole, another low pressure is bound to develop and it appears at around 600 north and south latitudes. It happens so, because the pole is excessively cold. It is quite natural to be a high pressure around poles. Therefore, two highs – subtropical and polar, the low pressure development is quite obvious. At this low pressure belt, rising air reaches near to the tropopause and gets diverted towards pole as well as sub-tropical direction. Hence, cell created by the circular motion from sub-tropical high to sub polar low, again upward, diverted to sub-tropical high and finally sinking is known as Ferrel cell.

The mid-latitude cell or Ferrel cell is also termed as polar front cell. You might know that in the higher latitudes the temperature contrast is greater. Polar areas are occupied by cold air mass while on the southern areas warm air mass develops. When both of the air masses come closer and interact with each other, the fronts are created. At the ground surface or near the surface the winds are blowing from southwest to northeast. Generally, these are known as westerlies. But aloft air circulation is opposite to it.

Polar Cell

The polar cell develops between 60° and poles in both hemispheres. Polar areas are thermally induced high pressure belts as it is developed because of the excessively cold conditions. Temperature is very low, air is dry, intermolecular space of the air is very less and hence, the density is very high. This air has the tendency to lie near the surface. Along the surface, cold winds blow towards the sub-polar low as easterlies. From sub-polar low, it is lifted up due to convergence with westerlies at polar fronts. From near the tropopause, it is diverted towards pole and finally completes a circulationcalled polar cell.

In brief, the three-cell atmospheric circulation model is symmetrical in shape and character with seasonal asymmetrical appearance. The best symmetry is during equinox and the maximum asymmetry is during solstice.

Vertical Atmospheric Circulation

Earth has strong gravitational pull, as we all know very well. In spite of the gravitation pull, the question arises as to what causes the air to rise and what happens when the air moves upward. To get the answers of these two questions, you must have got some ideas from the discussion about different atmospheric cells. You have studied that the condensation and precipitation release latent heat and it generates additional energy to boost the diversion process from equatorial zone. You would study about the condensation and precipitation. Most of the condensation and precipitation processes are related to the vertical atmospheric circulation. Vertical rise of air is characterized by:

- Decrease in temperature with elevation,
- Decrease in density of air with elevation,
- Decrease in pressure with elevation,
- Increase in intermolecular space of the air with elevation,
- Decrease in temperature is due to expanding air mass,
- Temperature, density and air pressure are highly correlated.

The upward moving air mass is bound to get altered and modified. The modification is seen in terms of air temperature, density and pressure. Most of the time, if moisture is available, condensation and precipitation are the natural outcomes. But here, uplifting or vertical motion in the air is caused by:

- Convective lifting,
- Orographic lifting,
- Frontal/cyclonic lifting,
- Convergent lifting.

In all these four cases, rising of the air cools. Relative humidity increases. Water holding capacity of the mass of air reduces. Dew level reaches by cooling of the air. In the presence of hygroscopic nuclei water vapour turns into tiny droplets. Through collision and coalescence process, the water droplet becomes bigger and fall down in numerous forms of precipitation. In other words, the vertical motion in the air causes cooling of the rising air as well as precipitation.

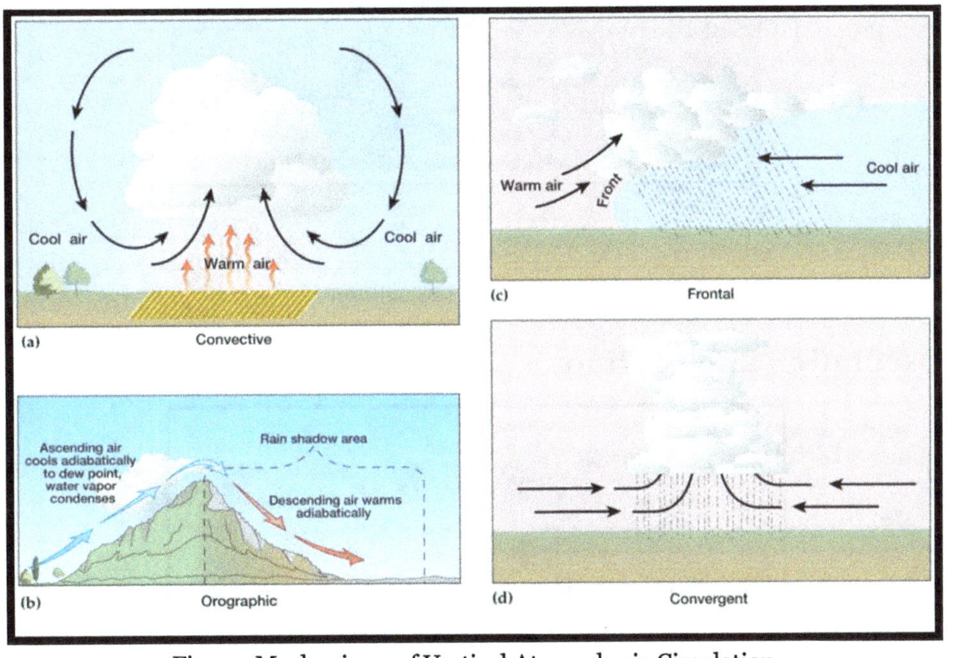

Figure: Mechanisms of Vertical Atmospheric Circulation.

Regional Vertical Atmospheric Circulation

Apart from the above mentioned vertical motion of the air, some other vertical circulations also occurat regional level, but they are not confined to an annual cycle. Their development is related to the regional synoptic changes. They are mainly:

- Walker circulation,
- Southern Oscillation and El Nino,
- Southern Oscillation and La Nina.

Walker Circulation

The concept of Walker Circulation was proposed by Sir Gilbert Walker. It is based on the difference in surface pressure and temperature over tropical western and eastern Pacific Ocean. Normally, tropical western Pacific is warm and humid with low pressure. Opposite to it, the tropical eastern Pacific is cool and dry with high pressure over the area. Therefore, in western part warm and humid airs rises and give rain and aloft it goes to eastern Pacific and sinks there and enhances the high pressure. This is the normal affair associated with general weather conditions.

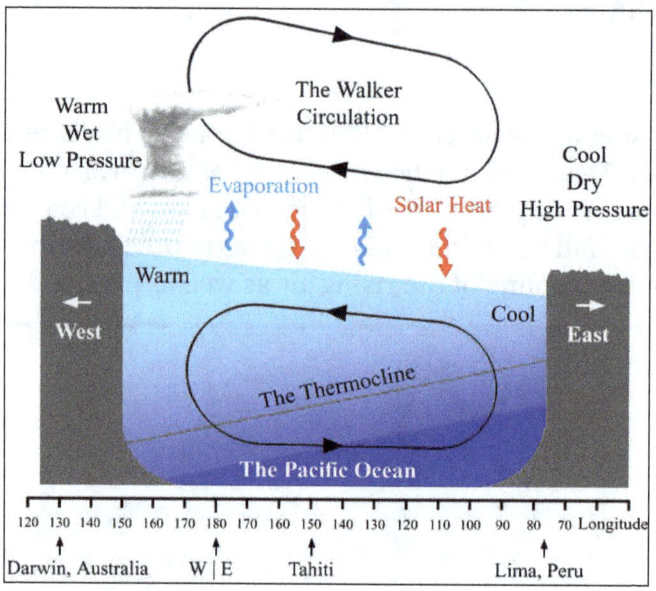

Figure: Walker Circulation.

Southern Oscillation and El Nino

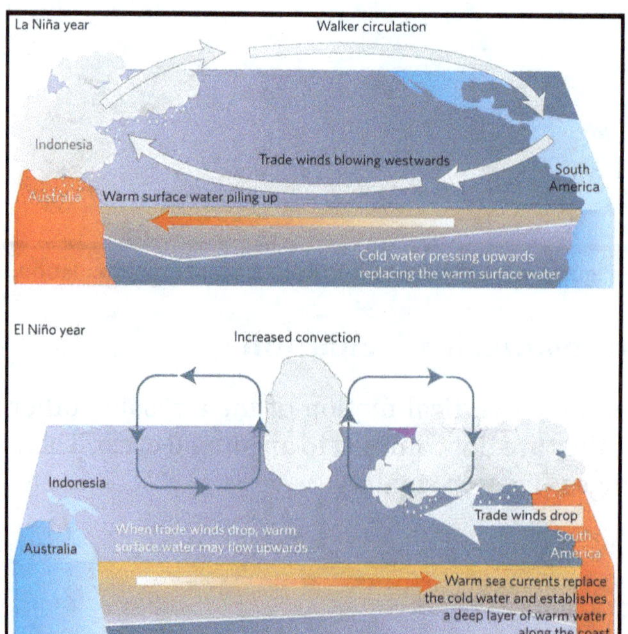

Figure: El Nino and La Nina.

This normal condition reverses after few years and it is termed as El Nino and Southern Oscillation. In this case, Walker Circulation weakens and warm water is spread to the eastern Pacific where low pressure develops. Warm and moist air rises, gives rain there and diverges towards west and subsides in the western Pacific. It is completely opposite to the Walker Circulation, and it is termed as El Nino.

Southern Oscillation and La Nina

When the western Pacific is excessively warm and humid and the low pressure is intense, it is called La Nina. La Nina is opposite to El Nino. Keeping the Walker Circulation in the center, both of these conditions are the two extremes of the events. All these are affecting the larger regional weather. These are also included in the vertical motion of the atmosphere.

Atmospheric Moisture

Humidity

The general term humidity refers to the amount of water vapour present in the air. The water vapour is gaseous form of water and its share is about 2 per cent in the composition of atmosphere. The content of humidity in air varies from place to place and time to time. It ranges from almost zero in the cold and dry air of arctic regions in winter to about 5 per cent in the hot and humid equatorial region. However, the total amount of water vapour remains nearly constant. The presence of water vapour has a significant role in determining the weather conditions of a particular place. Indeed, meteorologists and climatologists have recognized water vapour as vital component of atmosphere as far understanding of atmospheric processes in concerned. Presence of water in all the three forms of matter in biosphere is a unique gift of nature. Some of the several reasons of its great significance are following:

- The quantity of water vapour present in a given volume of air indicates the precipitation potential.
- Water vapour, as primary greenhouse gas, is a regulator of terrestrial heat.
- The presence of water vapour represents the amount of latent heat stored up in the atmo-sphere for the genesis of storms (cyclones)and turbulences.
- It determines the stability and instability of air masses.
- The amount of water vapour also influences human efficiency and health conditions.

Three States of Water

Water is the only substance on Earth that naturally occurs in three physical states: solid, liquid, and gas. Depending on temperature and atmospheric pressure, water can change from one state to another, a process called physical phase change. Because of this, some geographical regions of the world experience humidity, rain, snow, or even a combination of all three.

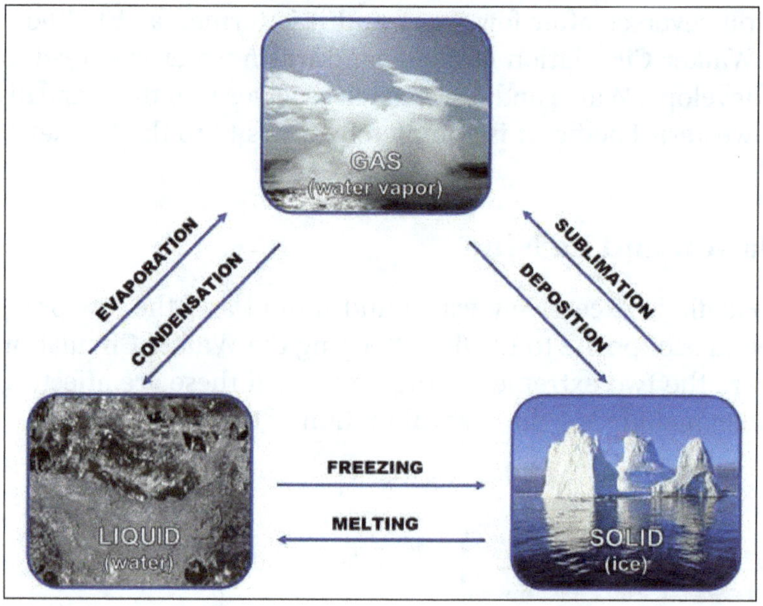

Phase Changes of Water. This figure demonstrates how the addition of heat can convert solid ice to liquid water through the process of melting, and with additional heat, the liquid water can be converted into gas (water vapor). Likewise, cooling converts liquid water to solid ice through the process of freezing. In general, moving up the diagram from bottom to top indicates adding heat/ energy, while moving from top to bottom the diagram indicates releasing heat/energy. Moving between gas and solid directly takes more energy than moving from solid to liquid to gas.

Hydrogen bonds are again the key. The number of bonds between molecules determines whether water will be a solid, liquid, or gas. In the solid state, water molecules have the maximum number of hydrogen bonds (4 per molecule), giving water the rigid characteristic of ice. In its liquid state, water has fewer hydrogen bonds, which accounts for its less-structured, fluid character.

As water changes from solid to liquid to gas, hydrogen bonds are broken, giving water molecules more freedom of movement. This cycling of water through its states is the basis for the hydrologic cycle that is essential for life on earth, purifying water and distributing it across land masses.

The Hydrological Cycle

The hydrological cycle predominantly involves the three processes – evaporation, condensation and precipitation. It is a gigantic system, energized by Sun, in which the vital link between lands and oceans is provided by atmosphere. Briefly, it represents the continuous movement of water from the oceans to the atmosphere, from atmosphere to the landmasses, from landmasses back to the oceans. Therefore, hydrological cycle represents interchanges of both physical state as well as geographical positions. Indeed, the movement of water in this complex hydrological cycle holds the key for distribution of moisture over the earth surface. Hydrological cycle is intricately related to all weather and climatic phenomena.

Water vapour in the atmosphere is derived through evaporation from the oceans and seas, terrestrial lakes, land water bodies (tanks and ponds), rivers, and ice fields and glaciers. These sources of moisture extend over about 75 per cent of the earth's surface. In addition to these sources, evaporation of soil moisture, transpiration from plants and animal respiration also contribute moisture to the atmosphere. Aridity, temperature and velocity of winds are the major determinants of the amount and intensity of evaporation. The relationship is direct, which means the high aridity, high temperature and high velocity of winds indicates higher rates and amount of evaporation. Evaporation from oceans is more as compared to continents. In continental areas maximum evaporation occurs between 10°N and 10° S latitudes. In case of oceans, the latitudinal zone of 10° -20° latitudes in the hemispheres records maximum evaporation. However, land-sea moisture exchanges and latitudinal moisture exchanges are common. The prevailing winds, air masses, cyclones, and other atmospheric disturbances play a very significant role in these moisture exchanges.

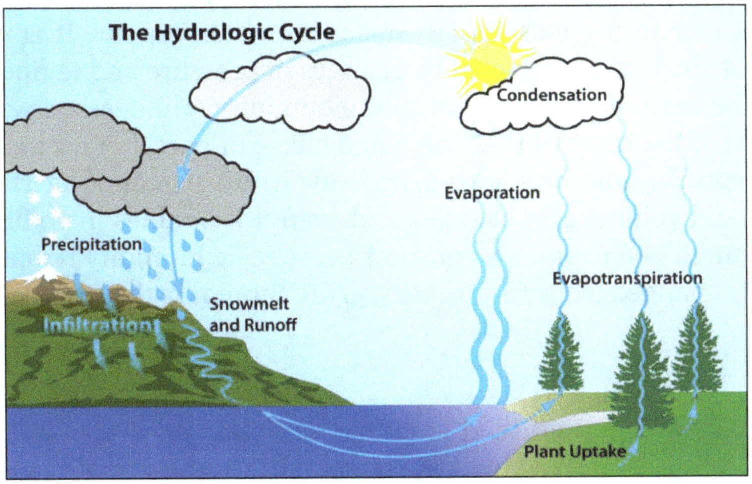

Figure: The Hydrological Cycle.

Humidity Measurements

Humidity is the general term which refers to the amount of water vapour present in the air. Several methods of measurement have been devised by scholars to express the presence of water vapour in air. The well-known measurements of humidity are following:

- Absolute humidity,
- Specific humidity,
- Mixing ratio,
- Vapour pressure and saturation,
- Relative humidity,
- Dew point.

Absolute Humidity

Absolute humidity is the mass of water vapour in a given volume of air. It is generally expressed as grams per cubic meter of air. Movement of air from one area to another, changes in temperature and pressure and results into changes in volume due to expansion or contraction. It changes the absolute humidity, even without addition or removal of moisture i.e. even when amount of moisture remains constant. For instance, if in one cubic meter parcel of air, at surface, mass of water vapour is 10 gram, absolute humidity is 10 g/m^3. If this parcel of air rises, due to decrease in pressure it expands and volume may become 2 m^3, then absolute humidity in the same parcel of air becomes 5 g/m^3, though there is no addition or removal of moisture. Therefore, it is difficult to monitor the water-vapour amount of a moving mass of air with the help of absolute humidity index. That is why it is not generally preferred by meteorologists.

Specific Humidity

The actual amount of moisture present in air is expressed by specific humidity. It is the ratio of mass of the water vapour to the mass of air, including water vapour. It is usually expressed as grams per kilogram of air. It is not affected by changes in pressure and temperature. To checks it, let us come back to the example quoted in absolute humidity. The mass of water vapour in a parcel of air was 10 gram, if the mass of that one cubic meter of parcel of air including water vapour is one kilogram the specific humidity is 10 g/kg. Now if this parcel of air rises, its pressure and temperature change and volume also changes and even if it doubles in volume the ratio of mass of water vapour and mass of air remains constant i.e. specific humidity remains 10 g/kg, whereas the absolute humidity changed from 10 g/m^3 to 5 g/m^3. Similar is the situation with mixing ratio.

Mixing Ratio

It is closely related to specific humidity. It is ratio of the mass of water vapour in a unit air compared to the remaining mass of dry air. It is measured in the units of mass, usually grams per kilogram, like specific humidity. It is not affected by changes in temperature or pressure. A value of 10 gram per kilogram as mixing ratio means the total mass for the mixture is 1010 g. In case of specific humidity of 10 gram per kilogram the total mass is one kilogram. These differences in specific humidity and mixing ratio, in most situations, are insignificant. However, it is not easy to determine absolute humidity, specific humidity and mixing ration by direct sampling. Therefore, other measurements like vapour pressure, relative humidity and dew point are used to express the moisture content of air.

Vapour Pressure and Saturation

This measure of the moisture content of the air is derived from the pressure exerted by water vapour. It represents the partial pressure exerted by water vapour in the air. It is measured in the same units used for barometric pressure i.e. milli bars. Like other atmospheric gases, water vapour exerts pressure. Due to evaporation, water molecules are added in the air above and it is reflected by small increase of air pressure. This increased pressure represents vapour pressure and is defined as that part of the total atmosphere pressure which is attributable to its water vapour content. However, with the passage of time as more and more molecules evaporate from the water surface, the gradually increasing vapour pressure in the air above forces more and more of these escaping molecules to return to the liquid. Finally at one stage, the number of water vapour molecules leaving and returning to the evaporating surface becomes same. This condition represents saturation and this time pressure is called saturated vapour pressure. The temperature of air at this time represents its dew point temperature, or frost point (if below freezing). Saturation vapour pressure varies with temperature and at temperatures below oC it is slightly lower over ice than over liquid water. Therefore, if temperature of the air above increases than the amount of moisture required for saturation will be more. This table represents the amount of water vapour required for saturation of one kilogram of dry air at different temperatures.

Table: Water Vapour Capacity (at mean sea-level pressure).

Temperature (°C)	Saturation mixing ratio (g/kg)
-40	0.10
-30	0.30
-20	0.75
-10	2.00
0	3.50
5	5.00
10	7.00
15	10.00
20	14.00
25	20.00
30	26.50
35	35.00
40	47.00

The Table shows that the saturation vapour pressure is temperature-dependent. At higher temperatures more water vapour is required for saturation. It is noteworthy that for every 10°C increase in temperature, the amount of water vapour needed for saturation almost doubles. Thus, four times more water vapour is needed to saturate one kilogram air at 25°C air than at 5°C. The difference between the water vapour holding capacity and actual amount of moisture present in air is known as the saturation deficit.

Relative Humidity

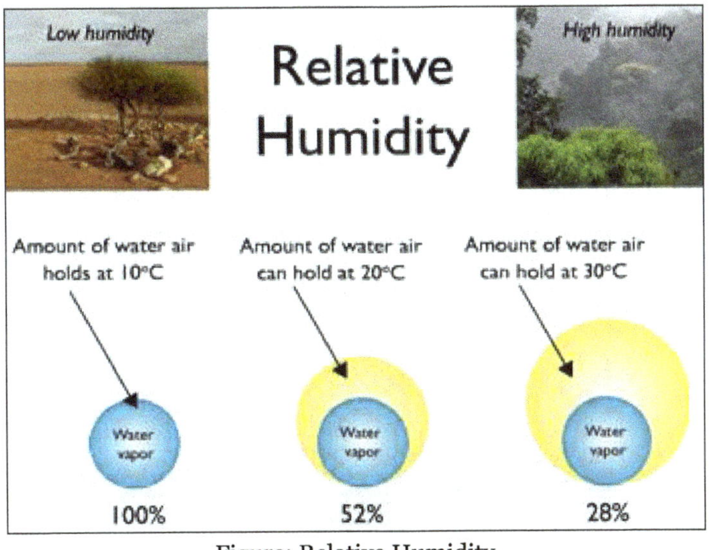

Figure: Relative Humidity.

Relative humidity is defined as the ratio of the actual water vapour content in the air to the amount of water vapour required for saturation at that particular temperature in the same air. Therefore, it does not represent the actual amount of water vapour in air but shows only how far or how near the air is from saturation. It has no unit of measurement and is always expressed in percentages. For example, in the Table, at 20 °C temperature one kilogram air gets saturated when it contains 14 g of moisture. Thus, if that air has presence of 7 g water vapour, the relative humidity is expressed as 7 (actual)/14 (saturation capacity)and multiplied by 100 gives the value in percentage, or 50 per cent.

The relative humidity can be changed in two ways:

- By addition or removal of water vapour.
- By change in temperature.

In our example, if through evaporation 7 g additional water vapour gets acquired at the same temperature (20 °C) by that parcel of air than relative humidity becomes 100 per cent. Likewise, relative humidity can become 100 per cent even without addition of water vapour by reducing the temperature to 10 °C. The air with 100 per cent relative humidity is known as saturated air. In atmosphere, the change in temperature and associated changes in relative humidity play a significant role in determining weather conditions. These changes include diurnal, seasonal and annual changes in temperature, temperature changes due to horizontal motion of air, temperature changes in fronts and temperature changes associated with the vertical movement of air. Relative humidity can also be expressed as the ratio of the observed vapour pressure to that required for saturation at a particular temperature. As follow:

$$RH = e/es$$

Where, RH stands for relative humidity, 'e' for vapour pressure, and 'es' for saturation vapour pressure. To derive values in percentage it is multiplied by 100.

Dew-Point Temperature

Dew Point Comfort Ranges	
less than 50	Very Dry
50-55	Comfortable
56-60	Pleasant
61-65	Slightly Humid
66-70	Humid
71-75	Very Humid
76+	Oppressive

Figure: Dew Point Based Comfort Ranges.

Dew-point temperature or simply dew point is another significant measure of humidity. It represents the temperature to which air is cooled to achieve saturation. You can recall the example quoted in relative humidity, that at 20 °C if the water vapour content is 7 g, then the relative humidity is 50 per cent. If temperature is reduced to 10 °C the air becomes saturated, with 100 per cent relative humidity. Therefore, 10 °C is the dew point or dew-point temperature. As it is a measure of the actual moisture present in a column of air, it is better than concept of relative humidity which shows only how far or how near the air is from saturation. It is easy to determine because of its direct association to the amount of water vapour in the air. Dew-point temperature has emerged as one of the most significant measures of humidity. It is used to determine association of the amount of moisture with human comfort.

It has already been concluded that the saturation vapour pressure is temperature dependent. For every 10 °C rise in temperature, the content of water vapour required for saturation doubles. Thus, comparatively cold saturated air (5 °C) holds half the water vapour of saturated air having a temperature of 15 °C and one-fourth that of hot saturated air with a temperature of 25 °C. It shows that the low dew-point temperatures represent relatively dry air and high dew-point temperatures are indicator of moist air. Therefore, the dew-point temperature is a good measure of the amount of water vapour in the air and is commonly used in weather maps. Surface dew-point temperature (in °F) based map of North America effectively shows the distribution of moisture over the continent.

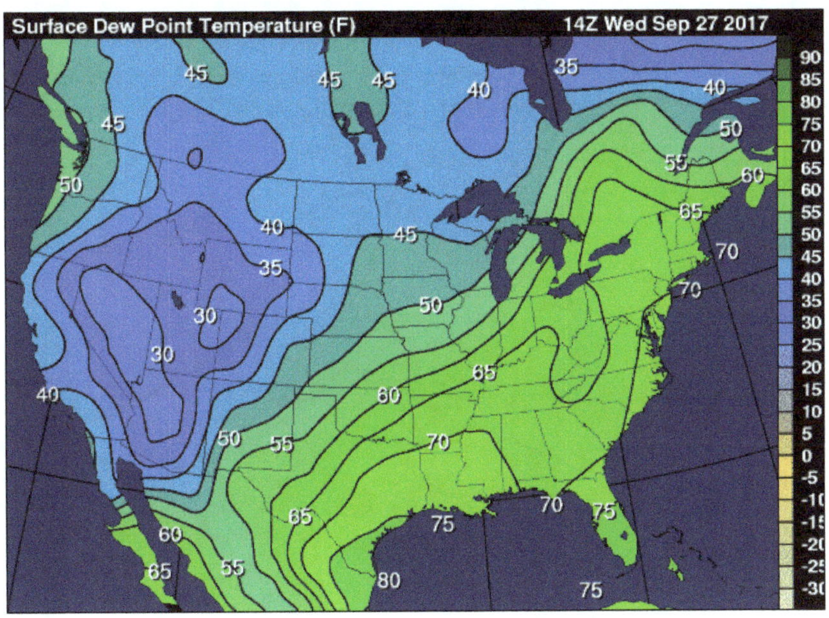

Figure: Dew Point Temperature in North America.

Distribution of Humidity

There are spatial and temporal variations in the distribution of humidity. These variations are common in vertical distribution as well as horizontal distribution. About half of the atmospheric moisture is concentrated below about 1450 meter and more than 90 per cent of the water vapour is below 5,575 meter. The seasonal variations in vertical distribution of humidity are mainly up to the height of 3,000 meter or so. It is noteworthy that the vertical concentration of humidity decreases form equator to poles.

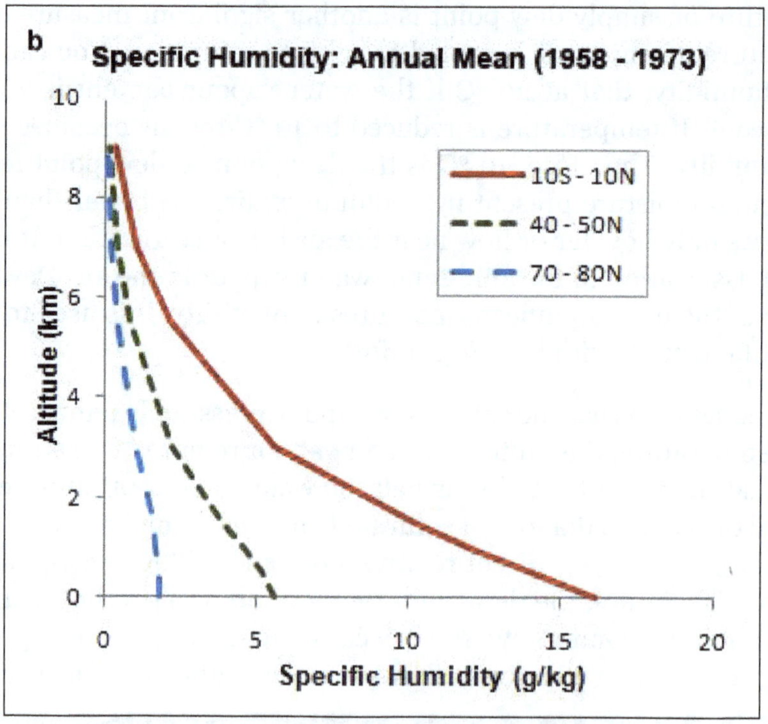

Figure: Vertical Variation of Atmospheric Water Vapour.

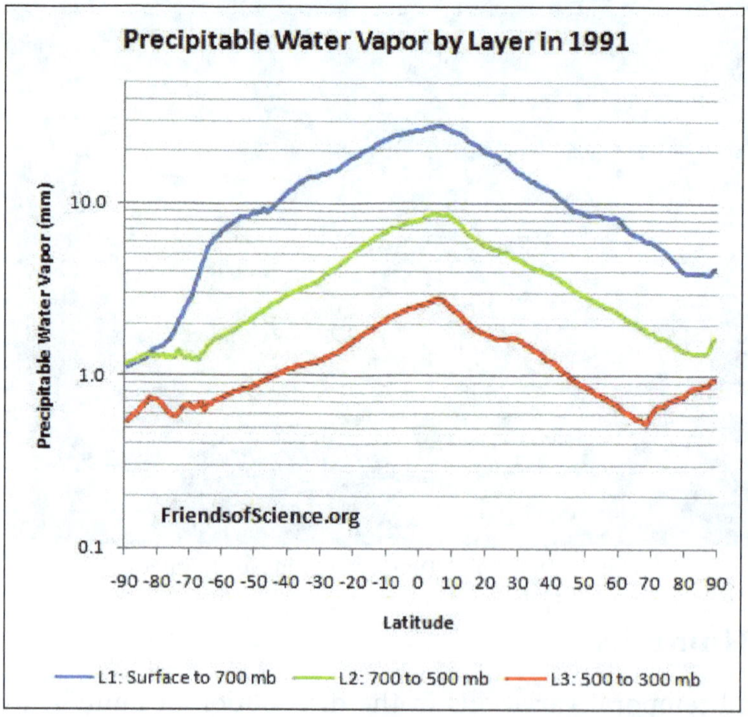

Figure: Distribution of Precipitable Atmospheric Water Vapour.

The mean storage of water vapour in the atmosphere is also known as the perceptible water content is about 2.5 cm. It is about 2.2 cm in January and 2.7 cm in July. The figure shows that perceptible water vapour decreases from equator to poles and the amount also decreases with height. In the first layer, which extends from surface to 700 mb i.e. up to the height of about 3 km it is in the range of about 2 mm in polar areas to about 12 mm over equator. The second layer, 700 to 500 mb means 3 km to 5.5 km height range has perceptible water with same pattern in the range of 1 to 9 mm. In the top layer, at height above 5.5 km the moisture content is limited and it is the range of 0.5 to 2 mm. Therefore, broad generalization about water vapour distribution is that it decreases with increase in altitude. The second measure generalization is that it decreases from equator to poles. This pattern is reflection of its direct relationship with temperature. The values are especially low over the deserts and high in equatorial and monsoon regions.

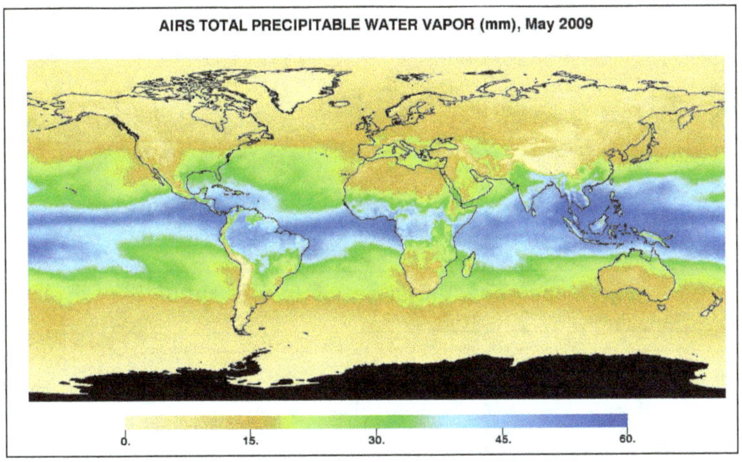

Figure: World Distribution of Precipitable Water Vapour.

The distribution of absolute humidity is extremely variable. It is maximum in the equatorial zone throughout the year and it is lowest in subtropical high pressure areas of Central Asia in winter and over the Antarctic. It decreases from equator towards poles. It also decreases from oceans to continental interiors.

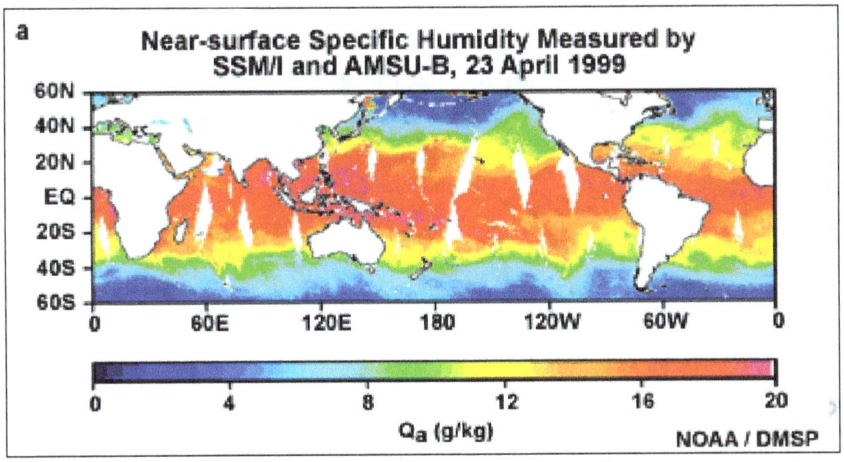

Figure: Distribution of Specific Humidity.

Specific humidity decreases from equator to poles. It is extremely low (about 0.2 g/kg) in cold and dry regions over arctic region in winter season and it is highest in equatorial region (about 18 g/kg). On global scale, the range of specific humidity varies from 100 to 200 times for largest values as compared to the least. It is considered as a geographer's yardstick of basic natural resource – water. Specific humidity is temperature controlled. Therefore, it is higher in summers than in winters. It is also higher over the oceans as compared to landmasses.

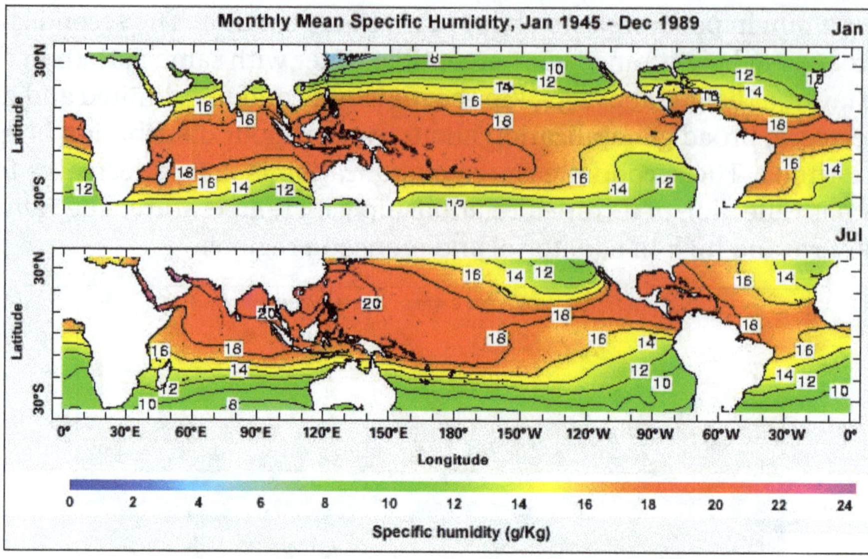

Figure: Seasonal Variations of Specific Humidity.

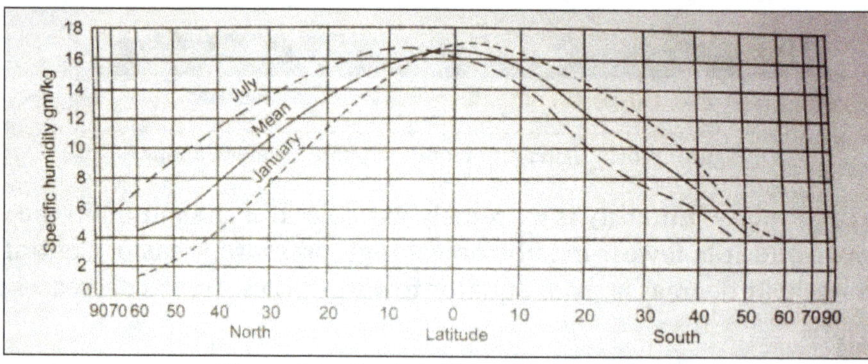

Figure: Distribution of Specific Humidity.

The vapour pressure distribution is similar to specific humidity pattern.

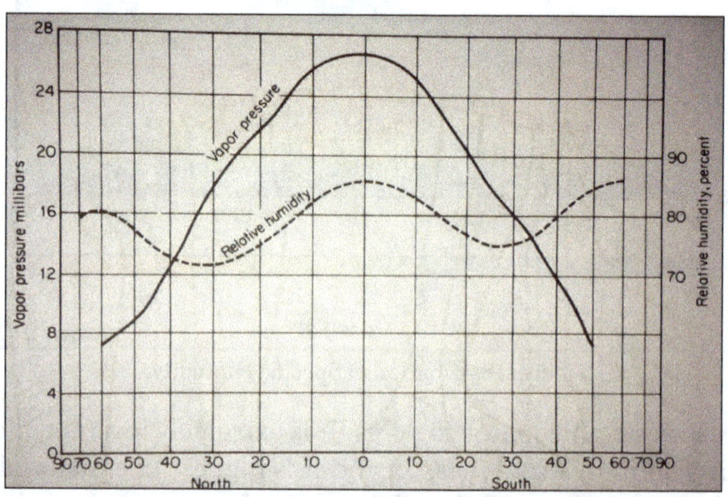

Figure: Distribution of Vapour Pressure and Relative Humidity.

It is noteworthy that the distribution pattern of relative humidity is different from specific humidity and vapour pressure. It is highest in equatorial zone and minimums in sub-tropical high pressure belts. From sub-tropical belts pole ward it again increases towards the waster lays wind belts. Another significant feature of the distribution of relative humidity is shifting of the belts of highest and lowest relative humidity with changes in overhead position of sun. Relative humidity also generally decreases from oceans to continental interiors.

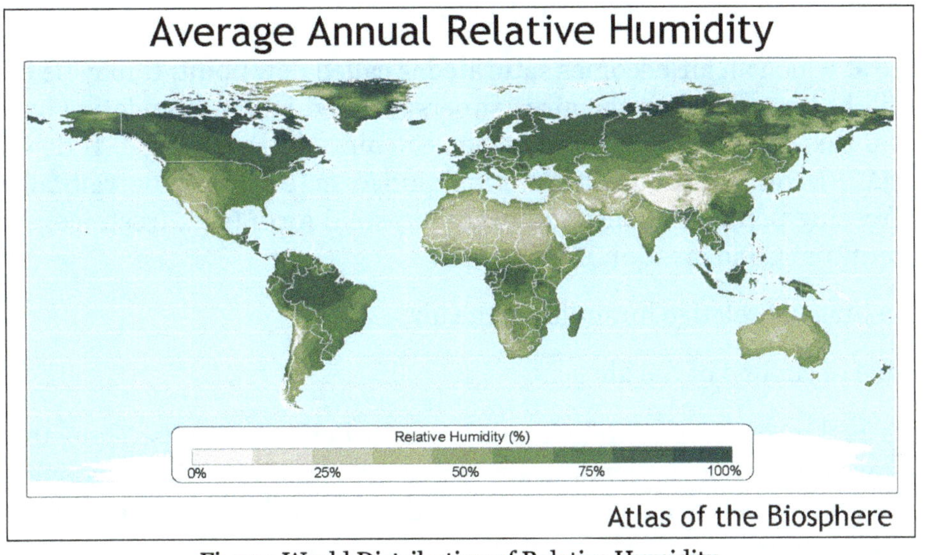

Figure: World Distribution of Relative Humidity.

There are latitudinal variations in the seasonal distribution of relative humidity. In the tropical regions the mean relative humidity remains higher in summer than in winter. Here, generally summers are wet and winter dry. In contrast, the high latitudes record the maximum relative humidity during humid winters. Remember that relative humidity on landmasses is dominantly controlled by temperature. Therefore, on the landmasses maximum relative humidity is recorded in winters as compared to summers. The exception to this generalization is monsoon region, where relative humidity is higher in wetter summer as compared with drier winter.

The diurnal variation in relative humidity is also well recorded. It is maximum in the early morning hours and minimum during the mid-afternoon time. This is due to indirect relationship with temperature.

Condensation

The transformation of gaseous form of water (i.e., water vapour) into solid form (ice) and liquid form (water) is called condensation. In other words, the process of change of water vapour into liquid form is called condensation. It is evident that condensation is opposite to evaporation. The process and mechanism of condensation depends on the amount of relative humidity present in the air. The air having 100 per cent relative humidity is called saturated air. An air may become saturated in two ways e.g., either:

- The absolute humidity at a given temperature is raised to equal the humidity retaining capacity of the air.
- The temperature of the air is reduced to such an extent that the humidity capacity becomes equal to its absolute humidity.

For example, the humidity capacity and absolute humidity of an air with 60 °F temperature are 5.7 grains and 4.1 grains per cubic foot respectively (relative humidity being 72 per cent), if the air is cooled to 50 °F temperature, the humidity capacity decreases to 4.1 grains which is equal to absolute humidity of 4.1 grains per cubic foot, thus the air becomes saturated as relative humidity becomes 100 percent, and hence condensation begins.

The temperature at which an air becomes saturated is called dew point. It may be pointed out that condensation will begin only when the air is supersaturated i.e., if the relative humidity exceeds 100 per cent, and this can be achieved only when the air is further cooled. If dew point is above freezing point (32 °F), condensation will occur in liquid form (e.g., dew, fog, rainfall etc.) but if dew point is below freezing point, condensation occurs in solid form (e.g., frost, ice, snow, hailstorm etc.).It is Apparent that Condensation Depends on:

- The percentage of relative humidity of the air,
- The degree of cooling of the air.

The air becomes cool when it rises while it gets heated when it descends. Thus, the ascending air may bring moist weather while descending air causes dry condition. Much cooling of the air is required in hot arid regions before dew point is reached. On the other hand, very little cooling causes condensation in humid regions. The heat released at the time of condensation is called latent heat of condensation.

Processes of Condensation

Some kind of cooling process is generally required to initiate and sustain condensation or sublimation. In general, the cooling is associated with four different processes:

- Cooling by expansion,
- Cooling by conduction of heat from the air,
- Cooling by radiation from the air,
- Cooling by mixing of warm and moist air with cold air.

These cooling processes are classified into two broad categories – adiabatic and nonadiabatic or diabatic processes. The first represents cooling by expansion and nonadiabatic class includes the rest three processes of cooling. The cooling processes are as follows:

Adiabatic Processes

Rising and expanding air, resulting into clouds, because of:

- Convection,
- Convergence of winds,
- Convergence of air masses,
- Orographic lifting.

A decrease in barometric pressure at surface may result into fog.

Nonadiabatic or Diabatic Processes

- Cooling due to loss of heat by radiation. Direct radiation from moist air may result into fog or clouds.
- Contact cooling with cold surface (conduction). Dew, frost, or fog may form.
- Contact cooling associated with the movement of air across a cold surface (advection).
- Mixing of warm air with colder air. It depends on sufficient moisture and cooling upto dew point. Fog and clouds will develop on hygroscopic nuclei.

Adiabatic Processes

Of all the types of cooling listed above, the most significant for condensation in free air is cooling due to rising air. The rising air experiences a drop in temperature, because internal energy is used in expansion caused by declining environmental pressure with altitude. As the individual molecules of gas are more widely diffused, they do not move so fast, the sensible temperature of the expanding gas declines. Conversely, descending air experiences an increase in pressure and temperature. These temperature changes without any heat exchange with surrounding, means without any heat being added or subtracted from the air are called adiabatic.

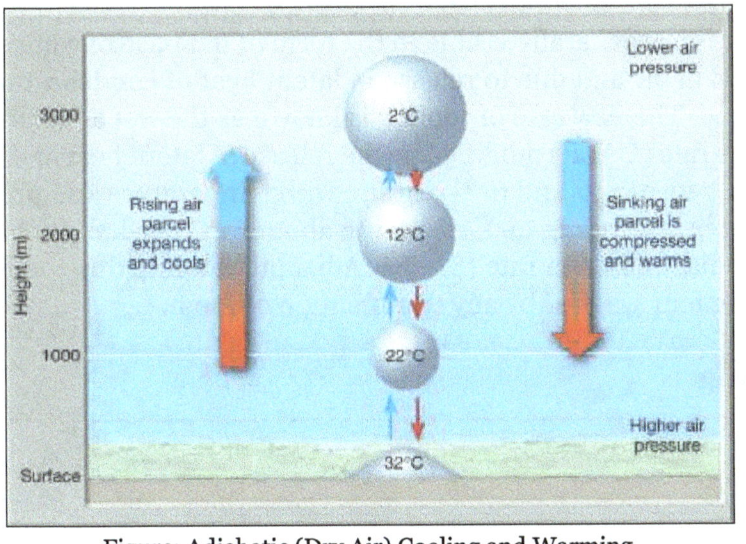

Figure: Adiabatic (Dry Air) Cooling and Warming.

Within the rising parcel of air the rate of fall of temperature is known as the dry adiabatic lapse rate and as shown in Figure, it is 10 °C per km of vertical uplift. This rate is effective till there is no condensation. The dew point also decreases gradually with air moving upward; the rate is 1.8 °C per km.

Let us assume that a parcel of air near the earth surface has a temperature of 20 °C and that here its dew point temperature is 11.8 °C, as shown in Figure. If this parcel of air starts rising gradually, its air temperature decreases much faster (as rate is 10 °C/km) than its dew-point temperature (as rate is 1.8 °C/km). At one km altitude, the temperature of rising parcel of air due to adiabatic change becomes 10 °C and it achieves dew point temperature and gets saturated. Further rise results in condensation of water vapour into minute water droplets, and thus cloud formation takes place. In case, this dew point is below freezing point or air is super cooled, the ice crystals will develop on sublimation nuclei. The end result of whole process is the formation of clouds.

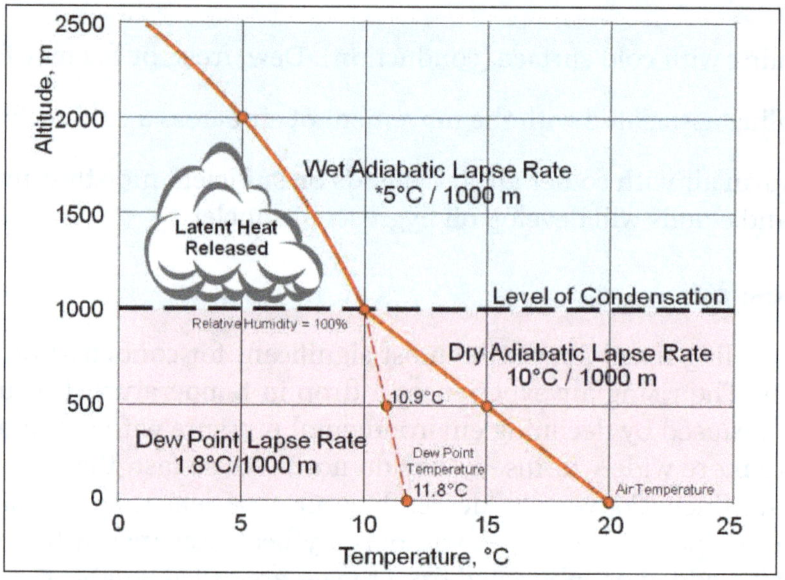

Figure: Adiabatic Change in Temperature in a Rising Parcel of Air.

The level of condensation is generally visible in the form of flat base of clouds. Condensation goes on in this rising parcel of air and due to release in latent heat of condensation; the adiabatic rate is substantially reduced. The new rate of cooling is known as the wet adiabatic rate; it is basically, the dry adiabatic lapse rate (DALR) modified by the release of latent heat of condensation. The wet adiabatic value varies between 3° and 6 °C per km, with temperature and pressure (and therefore with altitude attained by ascending air), averaging about 6 °C per km. Wet adiabatic rate is also known as saturated adiabatic lapse rate (SALR). Adiabatic cooling due to lifting is accomplished by any of the following four ways or by any combination of them.

Orographic Lifting

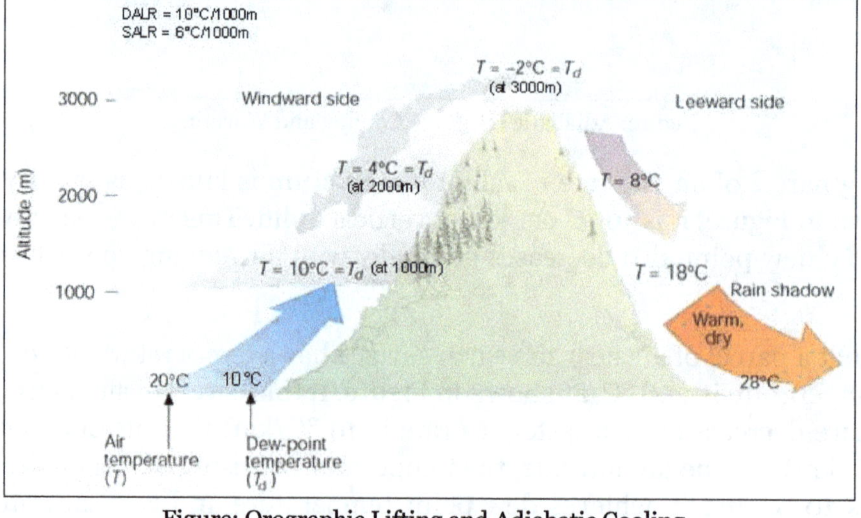

Figure: Orographic Lifting and Adiabatic Cooling.

It is ascension of air caused or intensified by any elevated terrain, such as mountains, hills and plateau acting as a barrier. Even a slight rise in terrain height, for instance on a coastal plain, can induce some rise of air. Even in this plain area the frictional drag of topography is more than that of the ocean surface. This frictional drag reduces the velocity of winds and air coming from sea is forced to rise over this congested air barrier in order to move towards continental interiors. In case of bigger barriers the lifting process is more intense. The rising air, through adiabatic cooling, often produces clouds and precipitation. On the leeward side the descending air adiabatically becomes warmer and dry and normally creates rain shadow zone.

Frontal Wedging

The convergence of warm and cold air masses results into formation of fronts. This weather phenomenon is very common in mid-latitudes. In frontal zone, the cooler and denser air works as barrier and over which warmer and less dense air rises. This process is known as frontal wedgingor frontal lifting. The collision of contrasting air masses forces results in condensation in fronts and associated temperate cyclones.

Figure: Frontal Wedging.

Convergence

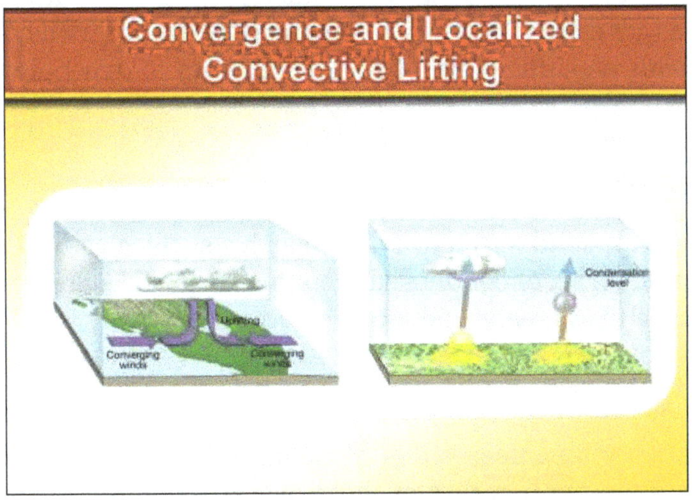

Figure: Convergence and Convective Lifting.

Frontal lifting is specifically related to convergence of air masses. In general, convergence takes place when winds of different directions or speeds meet one another. In the tropical areas, the converging winds have normally similar characteristics. For instance, converging winds at Florida peninsula from the Atlantic side as well as the Gulf of Mexico. Most important, convergence as a mechanism of forceful lifting is a major contributor to the weather conditions associated with tropical cyclones.

Localized Convective Lifting

Convection as process resulting to adiabatic cooling is normal whenever the earth's surface is warmer than the air above. In summer season, on warm days the differential rates of heating of surface results into formation of parcels of air warmer than surrounding air. This process results into localized convective lifting. This warm and moist air may rise up to level of condensation and result into clouds and light rain. This type of convection mechanism also contributes significantly to the lifting initiated by the other mechanisms. It should be noted that while the other mechanisms force air to rise, convective lifting occurs automatically because air is warmer and less dense than adjoining air.

Cooling by Lowering of Barometric Pressure

Adiabatic cooling occasionally can result into condensation due to lowering of barometric pressure in a moist air at the ground level, which occurs because of change of general distribution of pressure over that area. The valleys or basins filled with stagnant moist air are the most probable and ideal sites for these conditions as less dense air moves overhead. This may result into the formation of extremely rare barometric fog. This cooling process also intensifies other types of fog.

Nonadiabatic or Diabatic Processes

These processes involve exchange of heat with an external source. In nonadiabatic processes the dew point can be achieved through the processes such as - cooling due to loss of heat by radiation, contact cooling with cold surface (conduction) contact cooling associated with the movement of air across a cold surface (advection) and due to mixing of warm air with colder air. Condensation in these processes depends on sufficient moisture, cooling up to dew point and for fog and cloud formation condensation nuclei.

Forms of Condensation

Condensation takes place on solid surfaces on the earth's surface as well as in free air near and away from it. The common forms of condensed moisture on surface are – dew and frost, adjacent to the surface in air – fog and at greater heights – clouds.

Dew

During night time surface cooling occurs due to radiation from earth surface. Occasionally, the cooling is so intense that the temperature of the moist air touching the surface falls below dew point. It results into dew formation or water droplets formation on features such as – stones, blades of grass and leaves of plants. The favorable conditions required for this nonadiabatic cooling of surface by radiation are following – long nights, clear sky, calm and moist air.

Figure: Dew – Form of Condensation.

Frost

Normally, frost is mistaken as frozen dew. Frost occurs when the dew point of air drops below the freezing point. It means white frost is result of sublimation or crystallization or deposition that is direct transformation from water vapour to ice crystal. The favorable conditions for frost formation are the same as for dew only addition is that dew point is below freezing point.

Absolute Instability

Absolute instability represents continued vertical motion of the ascending parcel of air till its temperature is not equal to that of surrounding environment. This, another extreme condition occurs when the environmental lapse rate is greater than the dry adiabatic rate and wet adiabatic rates. In case there is general instability in the atmosphere, a force, whatsoever, is required to aloft a given parcel of air. The intense surface heating by solar radiation or presence of a topographic barrier in the path of air streams can provide such a mechanism for initial upward motion.

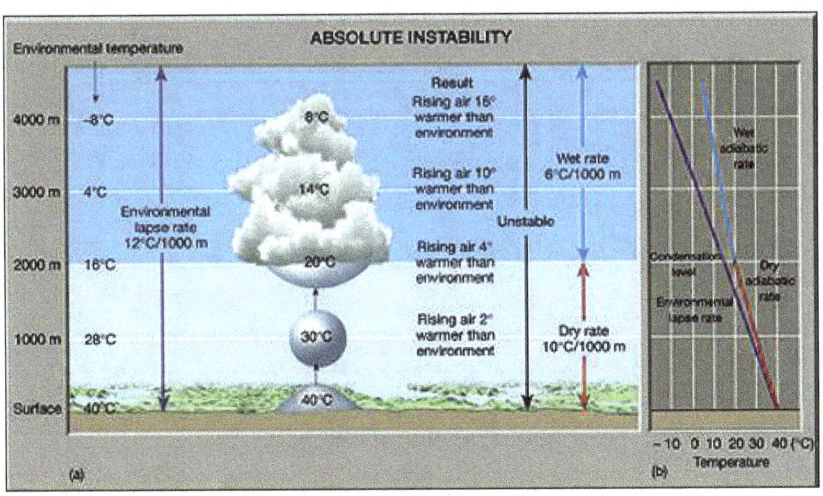

Figure: Absolute Instability.

Instability requires higher environmental lapse rates as compared to adiabatic rates. Assume that environmental lapse rate and wet adiabatic rates are 12 °C and 6 °C per km respectively, and you remember that dry adiabatic rate is fixed i.e. 10 °C per km. Assume, at surface temperature is 40 °C, and due to initial trigger-effect by some impulse the parcel of air starts moving upward. At 2 km altitude, the temperature in this parcel of air becomes 20°C and condensation starts because of saturation and dew point. The surrounding environment is cooler with 16°C temperature. As this parcel of air is warmer it will continue upward movement at wet adiabatic rate. At one km height from level of condensation this rising parcel of air is 10°C warmer than surrounding environment. The ascending parcel of air remains always warmer than its surrounding environment and will continue to move upward because of its own buoyancy. The end result will be vertical clouds and precipitation.

Conditional Instability

A more frequent type of atmospheric instability is known as conditional instability. This occurs when the moist air has an environmental lapse rate between the dry and wet adiabatic rates. It simply means that the atmosphere is stable with respect to unsaturated parcel of air, but unstable in relation to the saturated parcel of air. Let us assume the same surface temperature 40°C, same wet adiabatic rate 6°C per km and dry adiabatic rate is 10°C and in this situation environmental lapse rate is 8°C per km, between wet and dry. The air is stable in the lower layers, but if the air is forced to rise, due to surface heating or orographic barrier or convergence, until the level of free convection. At this level the rising parcel of air becomes warmer than its surroundings and is free to rise.

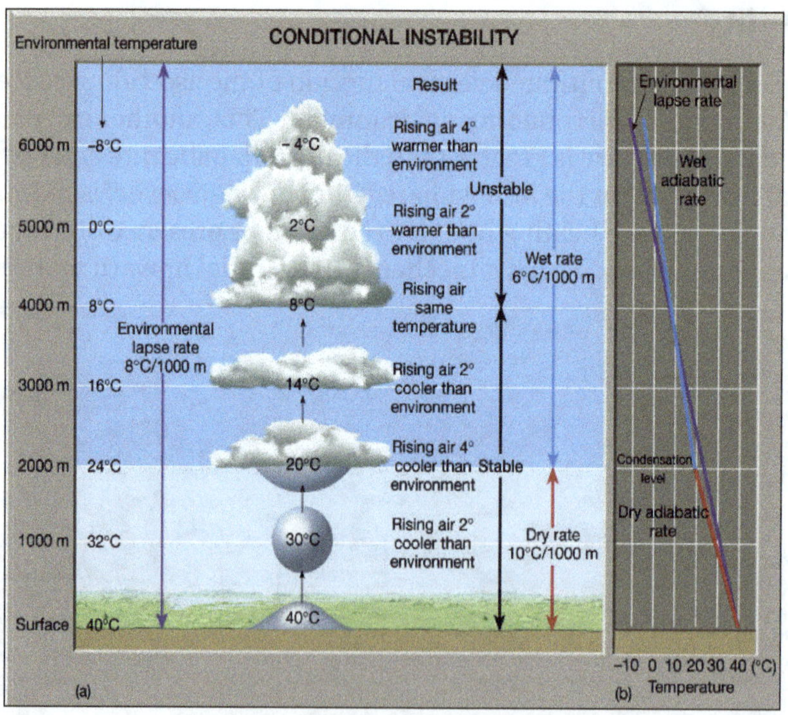

Figure: Conditional Instability.

In this example, at the level of condensation the rising air is 40C cooler than surrounding. It means it is stable, but if the lifting force continues and forces it above this level, wet adiabatic rate operates and at 4 km height the temperature of rising air becomes similar to surrounding. This represents the level of free convection, because this height onwards this parcel of air will move upward on self-basis due to release of latent heat of condensation and it will remain warmer than environment. From this level along its ascent, the parcel will continue to rise without an outside force.

This situation is called conditional instability, because instability dependents on the air parcel becoming saturated. The term "conditional" is used because the air must be forced upward before it reaches the level where it becomes unstable and rises on self-basis. Conditional instability is very common because the environmental lapse rate frequently prevails between the dry and wet adiabatic rates. The prerequisites for conditional instability are – the ascending parcel of air should be saturated, and it should get a very strong outside initial force to aloft it not only up to the height of level of condensation rather to the level of free convection.

Convectional Instability

So far in our discussion on, stability, absolute instability and conditional instability the situations were about a parcel of air uprising through a layer of air that was static vertically. But occasionally, there are conditions when an extensive layer (several hundreds of meters thick, extending over thousands of kilometers) is forced to up rise over an extensive orographic barrier or by some other mechanism. Within this extensive and thick layer of air the lower portion may be moist and upper portion may be dry.

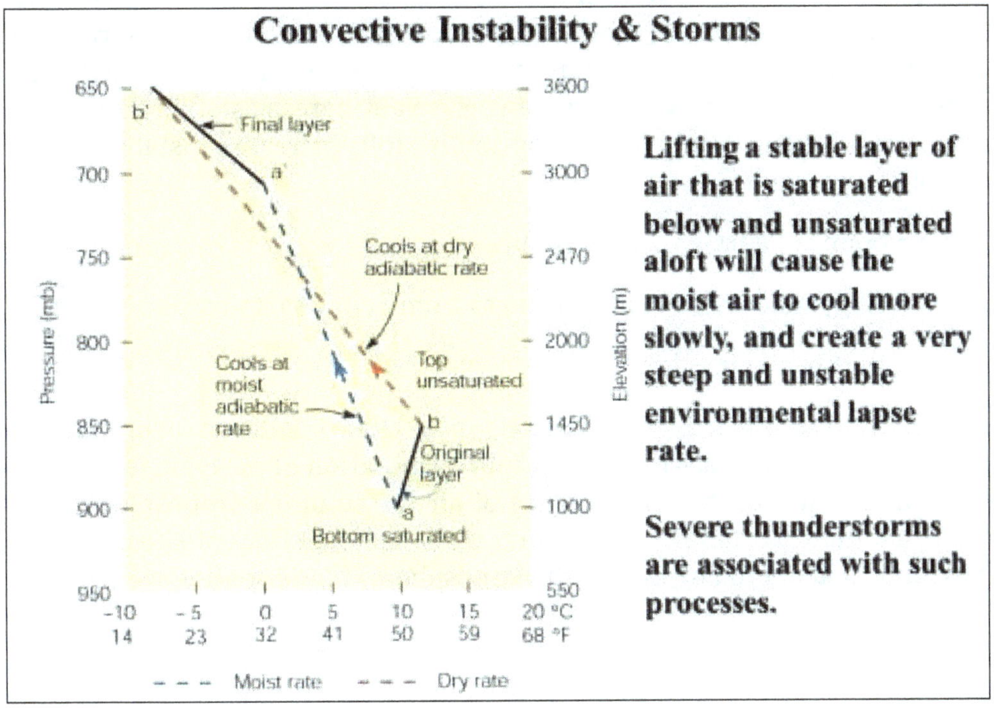

Figure: Convective Instability.

In case, the whole layer is forced for vertical motion, the drier air at 'b' cools at the dry adiabatic rate, and same is the case at 'a', till the time condensation level is not achieved by lower portion. After reaching condensation level the lower part air mass cools at the wet adiabatic rate. This process finally increases the actual lapse rate of the total thickness of the raised layer, reflected in the figure by change from ab to a'b'. Here, ab represents the initial lapse rate (stable) and a'b' represents final lapse rate (unstable) which is more than the saturated adiabatic lapse rate and that is why this line is steeper and it shows instability. Finally the layer becomes unstable and may overturn, as lower part (a) ascends faster than upper part (b). It is called convective or potential instability. Hence, it may be concluded that, in general, saturation lifting produces more instability than unsaturated lifting. The process of convective instability is vital in the dynamics of weather. When convectively or conditionally unstable air masses with conditions mentioned above are lifted, they produce vertical clouds, cumulonimbus and result in heavy precipitation with thunder and lightning. Further, convective instability also plays role in the formation of cyclones and thunderstorms.

Mechanical Instability

Sometimes, abnormal conditions prevail and lapse rates in upper layers of troposphere become too high, lapse rates of 20 °C to 35 °C per km. This makes upper layers denser and heavier as compared to underlying layers. In case of such extraordinary conditions there is an automatic overturning of air without any initial trigger. This kind of situation represents mechanical instability. It contributes in the genesis of tornadoes, the most violent revolving storms.

Changes in Stability

It can be concluded that unstable parcel of air due to its own buoyancy ascends freely and on the other hand, stable air resists vertical motion. The most stable conditions are represented by the inversion of temperature. But steeper environmental lapse rates are associated with instability. Therefore, those factors which cause cooling of surface air increase stability and those which enhance warming effect in surface air in relation to air aloft, increases instability. Stability is enhanced by the following conditions:

- Radiation cooling of surface during night.
- Contact cooling (conduction) of an air mass from below as it passes over a cold surface.
- Subsidence within an air column.

Subsidence results into substantial modifications in weather conditions. For example, a subsidence of about 300 m in an air mass may result into evaporation of all average size cloud droplets. The subsidence, due to horizontal convergence of air in the upper troposphere associated with high pressure belts, induces stability. Therefore, the warming effect of subsidence is enough to evaporate the clouds present in any part of the atmosphere.

Instability Enhancement Processes

The instability is enhanced by following processes:

Intense Surface Heating and Convection

Nocturnal cooling and stability are replaced by surface heating on a clear sunny day. The stability is replaced by instability. Due to this heating the lower atmosphere often becomes warmed enough to generate vertical motions in parcels of air. This heating results into onset of convectional process. The warm and moist parcels of air are cooled adiabatically below the dew point. The rising parcel of air gets transformed into a cumulus cloud and its flat base shows the level of condensation. There are many possibilities, one situation can be that this parcel of air is cooled enough and it ceases to rise and condensation is no more operational. This may result into dissolving of cumulus cloud after some downdraft. However, under a different set of conditions, favoring instability, convection may manifest itself in the form of great vertical clouds, cumulonimbus and associated heavy precipitation and lightning and thunder. Another outcome may be a convection cyclone energized by the latent heat of condensation. In the equatorial low pressure belts intensive surface heating results into formation of convectional currents and daily in the afternoon heavy rainfall is caused by cumulonimbus clouds.

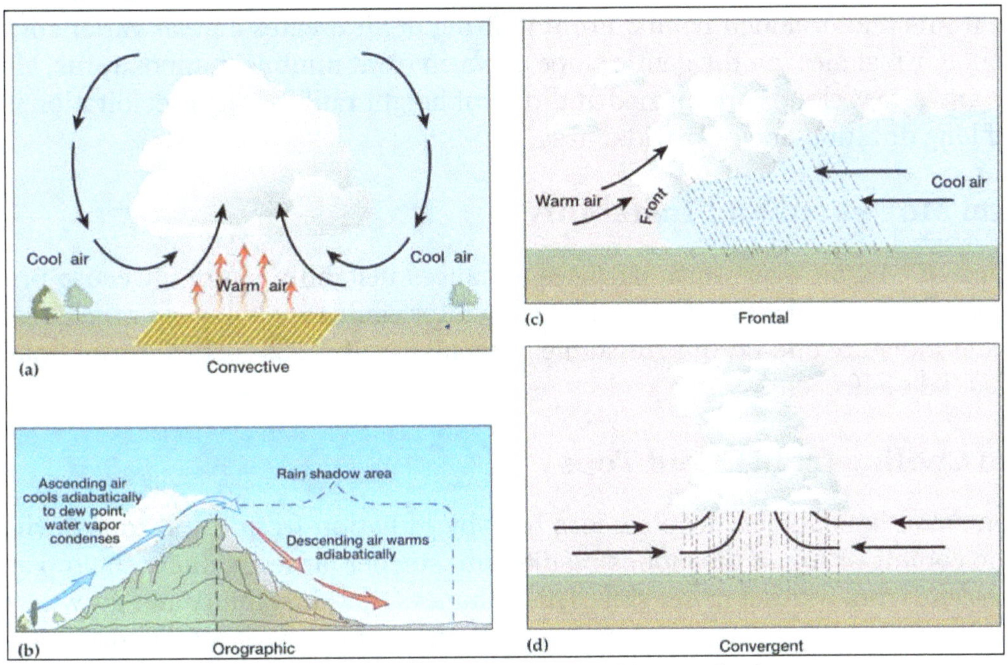

Figure: Atmospheric Instability and Lifting Mechanisms.

Lifting Mechanisms

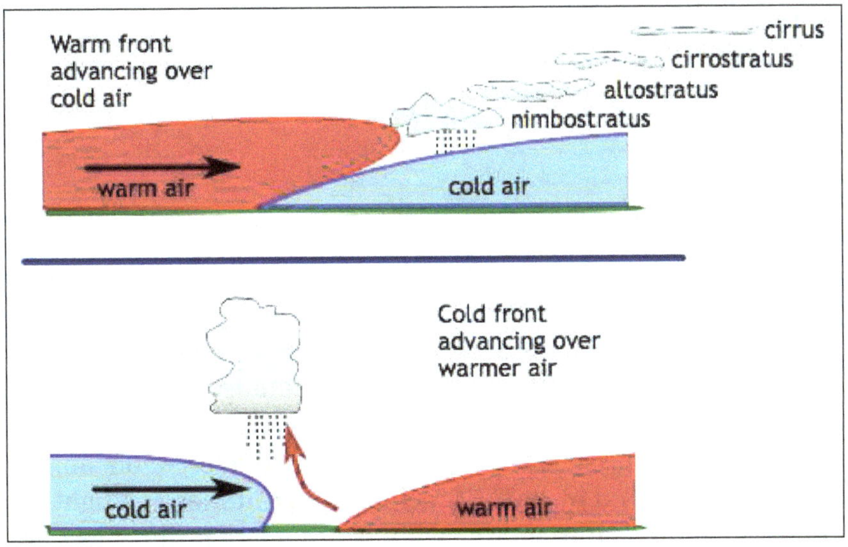

Figure: Fronts and Associated Weather.

Precipitation in substantial amount is induced by spontaneous convectional rise of warm and moist air and another mechanism enhancing instability is the forced rise of the moist air. The forced upward movement of air is caused by processes such as orographic lifting, frontal wedging, and convergence. All these lifting processes enhance instability and cause precipitation. The forced rise of moist air on the windward slope of orographic barrier results into adiabatic cooling, condensation and precipitation.

Convergence of winds results into lifting of air and enhances instability and causes precipitation. Convergence of two contrasting air masses results into instability along with formation of warm fronts, cold fronts and occluded fronts. Frontal lifting of air creates a great variety of clouds and precipitation. For instance, on the gentle slope of warm front nimbus, nimbostratus, altocumulus, cirrostratus and cirrus clouds are formed at different height ranges. The precipitation is moderate to gentle, of long duration and over wider area.

Horizontal Movement and Instability

The advection of cold air over warm surfaces enhances instability and may cause precipitation. For instance, the polar cold and dry air masses in winter while traversing over the Great Lakes acquire heat and moisture and become unstable. This instability causes cloud formation and snowfall known as "lake-effect snows".

Radiation Cooling from Cloud Tops

Although on small scale, cloud droplets loss heat by radiation from cloud tops during evening, the radiation cooling at top of the clouds results into steeper lapse rate near the top and induces additional ascend from warmer lower side. This enhances their instability and growth. This mechanism is used to explain the nocturnal thunderstorms from clouds that ceased to grow prematurely at the end of day.

Ocean Currents

Warm ocean currents enhance instability and cold ocean currents enhance stability. For instance, majority deserts of the world are present in sub-tropical region mainly due to presence of high pressure belts, but stability and associated aridity is enhanced by cold ocean currents prevailing on western side of the continents.

Instability and Daily Weather

The daily weather conditions are normally determined by stability and instability operating in the atmosphere. The calm and clear sky conditions represent atmospheric stability. However, ascend of stable air, due to forced mechanism, results into formation of clouds which are fairly widespread but have limited vertical extent. Precipitation from such clouds, if any, is invariably light. Light drizzle and overcast sky indicate the forced uplift of the stable air. On the contrary, cauliflower-shaped cumulonimbus clouds associated with the convective unstable air currents have great vertical extent and are accompanied by heavy precipitation, thunder and lightning. Therefore, the type of clouds and nature of precipitation are symbols of type of atmospheric stability and instability.

On hot summer afternoons, due to intense surface heating, instability intensifies and, parcels of air get heated and through convection process move upward. In case, they are moist enough and able to rise to the level of condensation, clouds develop to give occasional mid afternoon showers. This precipitation is light and of short duration. Precipitation itself cools the surface and breaks the convectional cycle and secondly, surface heating of this type provides limited instability. But in equatorial low pressure belts surface heating (convection), convergence of winds and upper air divergence of Hadley cell intensify the level of instability. The result is heavy precipitation from cumulonimbus clouds extending up to tropopause with anvil head.

The inversion of temperature produces the most stable conditions in the atmosphere. In this situation, the air near surface becomes cooler and heavier than the air aloft. Radiation inversion, advection inversion; air drainage inversion and upper air inversion due to subsidence are common indicators of atmospheric stability. Long nights, clear sky, calm and dry air conditions are ideal for inversion of temperature. In winter, dew, frost and fog develop due to contact cooling. Radiation cooling, advection cooling and associated fogs represent stability. An extensive fog is guaranty of atmospheric stability. Subsidence can result in a temperature inversion aloft. Hence, the diurnal and seasonal characteristics of weather are strongly related to atmospheric stability and instability.

Instability and Distribution of Precipitation

In the distribution of the precipitation most important determinants are stability and instability of atmosphere. The global mean annual precipitation is about 97 cm. The areas of atmospheric instability receive very high precipitation; contrary to this the areas of atmospheric stability receive very low precipitation. In the latitudinal belts of atmospheric convergence due to adiabatic cooling cloud formation and precipitation are intense. The equatorial doldrums due to convergence and convection have instability and have precipitation maxima. The average annual precipitation in equatorial belt is about 175 – 200 cm, and majority precipitation is received in the afternoon and it occurs throughout the year. The second maxima of precipitation is in the mid latitude convergence zone. In this zone the instability is intensified by the convergence of warm moist tropical air masses with cold polar air masses in frontal zone. The instability in these belts is associated with convergence, frontal wedging and temperate cyclones and results into throughout the year precipitation. Here, the average annual precipitation is about 100 – 125 cm. On the contrary, the subtropical high pressure areas of subsidence and divergence do not encourage cloud formation and precipitation due to atmospheric stability. These are the most arid areas and are predominated by deserts, with average annual rainfall less than 25 cm. Likewise, the stable polar areas due to high pressure or subsidence receive less than 25 cm average annual precipitation and represent cold deserts.

Monsoon areas receive high rainfall due to atmospheric instability in summer season. The differential rates of heating of land and sea, resultant moist and warm onshore winds, orographic lifting and local convective aloft are the reasons for instability in this climatic region. In the world, wherever the coastal areas are flanked by mountain ranges the onshore ascending winds due to adiabatic cooling cause cloud formation and heavy precipitation on the windward side and rain shadow regions on leeward sides. It is because of instability produced by orographic lifting that mountain areas record the highest precipitation in the world, for instance, Meghalaya plateau, Western Ghats, western Rockies and Andes, eastern Great Dividing Range and Southern Alps.

The air masses and prevailing winds moving over relatively warmer areas become unstable and may result into clouds and precipitation. In case of precipitation associated with tropical disturbances such as cyclones, hurricanes, tornadoes, the upper air divergence, lower air convergence and convection and adiabatic lifting result into heavy precipitation, by cumulonimbus or thunder clouds. The warm ocean currents are conducive to greater precipitation because they cause and intensify instability. On the contrary, cold ocean currents produce stability and extensive fog in suitable situations. Hence, there is intricate relationship between stability and instability and world distribution of precipitation.

Precipitation

Precipitation is any liquid or frozen water that forms in the atmosphere and falls back to the Earth. It comes in many forms, like rain, sleet, and snow. Along with evaporation and condensation, precipitation is one of the three major parts of the global water cycle.

Precipitation forms in the clouds when water vapor condenses into bigger and bigger droplets of water. When the drops are heavy enough, they fall to the Earth. If a cloud is colder, like it would be at higher altitudes, the water droplets may freeze to form ice. These ice crystals then fall to the Earth as snow, hail, or rain, depending on the temperature within the cloud and at the Earth's surface. Most rain actually begins as snow high in the clouds. As the snowflakes fall through warmer air, they become raindrops.

Particles of dust or smoke in the atmosphere are essential for precipitation. These particles, called "condensation nuclei," provide a surface for water vapor to condense upon. This helps water droplets gather together and become large enough to fall to the Earth.

A common misconception is that when raindrops fall, they have a teardrop shape. In fact, smaller raindrops (ones that are approximately 1 millimeter (0.039 inches) across) are almost perfectly spherical. Larger raindrops (2–3 millimeters (.078-.118 inches) across) are also round, but with a small indent on their bottom side. They look more like kidney beans when falling to the Earth. Very large rain drops (larger than 4.5 millimeters (.177 inches)) have a huge indent and look more like a parachute. These extra-large drops usually end up splitting into two smaller droplets. The indents on raindrops are caused by air resistance.

Precipitation is always fresh water, even when the water originated from the ocean. This is because sea salt does not evaporate with water. However, in some cases, pollutants in the atmosphere can contaminate water droplets before they fall to the Earth. The precipitation that results from this is called acid rain. Acid rain does not harm humans directly, but it can make lakes and streams more acidic. This harms aquatic ecosystems because plants and animals often cannot adapt to the acidity.

References

- Significance-of-the-atmosphere, geophysical: courses.lumenlearning.com, Retrieved 07, March 2020
- Insolation_Earth's_Heat_Balance_Different_Atmospheric_Circulations_Global_Winds_Cyclones: visio-nias.in, Retrieved 12, April 2020
- Atmospheric-circulation, large-scale-weather: meteoblue.com, Retrieved 15, May 2020
- Waters-three-physical-states, water: healingearth.ijep.net, Retrieved 05, January 2020
- What-is-condensation-climatology-geography, climatology-2-2798: geographynotes.com, Retrieved 28, July 2020

Chapter 3

Subdisciplines of Climatology

Climatology can be divided into various branches; each corresponding to a specialized niche. Paleoclimatology, paleotempestology, historical climatology, tornado climatology, tropical cyclone rainfall climatology, urban climatology, dendroclimatology, etc. are a few branches of climatology examined in detail in this chapter.

Paleoclimatology

Paleoclimatology (in British spelling, palaeoclimatology) is the study of changes in climate taken on the scale of the entire history of Earth. It uses a variety of proxy methods from the Earth and life sciences to obtain data previously preserved within things such as rocks, sediments, ice sheets, tree rings, corals, shells and microfossils. It then uses the records to determine the past states of the Earth's various climate regions and its atmospheric system. Studies of past changes in the environment and biodiversity often reflect on the current situation, specifically the impact of climate on mass extinctions and biotic recovery.

History

The scientific study field of paleoclimate began to form in the early 19th century, when discoveries about glaciations and natural changes in Earth's past climate helped to understand the greenhouse effect.

Reconstructing Ancient Climates

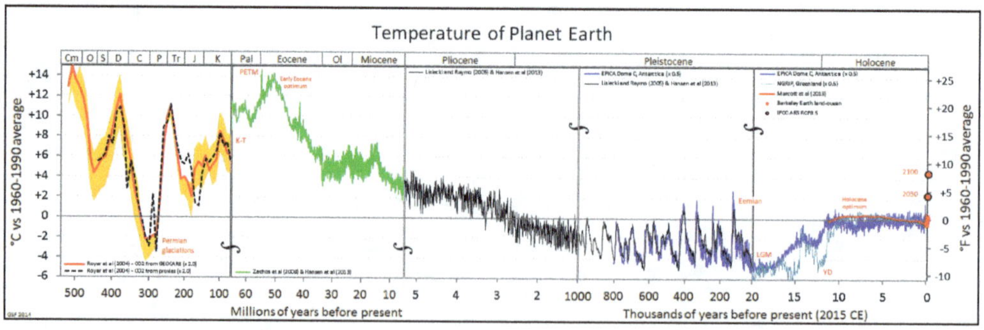

Palaeotemperature graphs compressed together.

Paleoclimatologists employ a wide variety of techniques to deduce ancient climates.

Ice

Mountain glaciers and the polar ice caps/ice sheets provide much data in paleoclimatology.

Ice-coring projects in the ice caps of Greenland and Antarctica have yielded data going back several hundred thousand years, over 800,000 years in the case of the EPICA project.

- Air trapped within fallen snow becomes encased in tiny bubbles as the snow is compressed into ice in the glacier under the weight of later years' snow. The trapped air has proven a tremendously valuable source for direct measurement of the composition of air from the time the ice was formed.
- Layering can be observed because of seasonal pauses in ice accumulation and can be used to establish chronology, associating specific depths of the core with ranges of time.
- Changes in the layering thickness can be used to determine changes in precipitation or temperature.
- Oxygen-18 quantity changes ($\delta^{18}O$) in ice layers represent changes in average ocean surface temperature. Water molecules containing the heavier O-18 evaporate at a higher temperature than water molecules containing the normal Oxygen-16 isotope. The ratio of O-18 to O-16 will be higher as temperature increases. It also depends on other factors such as the water's salinity and the volume of water locked up in ice sheets. Various cycles in those isotope ratios have been detected.
- Pollen has been observed in the ice cores and can be used to understand which plants were present as the layer formed. Pollen is produced in abundance and its distribution is typically well understood. A pollen count for a specific layer can be produced by observing the total amount of pollen categorized by type (shape) in a controlled sample of that layer. Changes in plant frequency over time can be plotted through statistical analysis of pollen counts in the core. Knowing which plants were present leads to an understanding of precipitation and temperature, and types of fauna present. Palynology includes the study of pollen for these purposes.
- Volcanic ash is contained in some layers, and can be used to establish the time of the layer's formation. Each volcanic event distributed ash with a unique set of properties (shape and color of particles, chemical signature). Establishing the ash's source will establish a range of time to associate with layer of ice.

Dendroclimatology

Climatic information can be obtained through an understanding of changes in tree growth. Generally, trees respond to changes in climatic variables by speeding up or slowing down growth, which in turn is generally reflected a greater or lesser thickness in growth rings. Different species, however, respond to changes in climatic variables in different ways. A tree-ring record is established by compiling information from many living trees in a specific area.

Older intact wood that has escaped decay can extend the time covered by the record by matching the ring depth changes to contemporary specimens. By using that method, some areas have tree-ring records dating back a few thousand years. Older wood not connected to a contemporary record can be dated generally with radiocarbon techniques. A tree-ring record can be used to produce information regarding precipitation, temperature, hydrology, and fire corresponding to a particular area.

On a longer time scale, geologists must refer to the sedimentary record for data.

Sedimentary Content

- Sediments, sometimes lithified to form rock, may contain remnants of preserved vegetation, animals, plankton or pollen, which may be characteristic of certain climatic zones.
- Biomarker molecules such as the alkenones may yield information about their temperature of formation.
- Chemical signatures, particularly Mg/Ca ratio of calcite in Foraminifera tests, can be used to reconstruct past temperature.
- Isotopic ratios can provide further information. Specifically, the $\delta^{18}O$ record responds to changes in temperature and ice volume, and the $\delta^{13}C$ record reflects a range of factors, which are often difficult to disentangle.

Sea floor core sample labelled to identify the exact spot on the sea floor where the sample was taken. Slight variations in location can make a significant difference in the chemical and biological composition of the sediment sample.

Sedimentary Facies

On a longer time scale, the rock record may show signs of sea level rise and fall, and features such as "fossilised" sand dunes can be identified. Scientists can get a grasp of long term climate by studying sedimentary rock going back billions of years. The division of earth history into separate periods is largely based on visible changes in sedimentary rock layers that demarcate major changes in conditions. Often, they include major shifts in climate.

Sclerochronology

Coral "rings" are similar to tree rings except that they respond to different things, such as the water temperature, freshwater influx, pH changes, and wave action. From there, certain equipment can be used to derive the sea surface temperature and water salinity from the past few centuries. The $\delta^{18}O$ of coralline red algae provides a useful proxy of the combined sea surface temperature and sea surface salinity at high latitudes and the tropics, where many traditional techniques are limited.

Limitations

A multinational consortium, the European Project for Ice Coring in Antarctica (EPICA), has drilled an ice core in Dome C on the East Antarctic ice sheet and retrieved ice from roughly 800,000 years ago. The international ice core community has, under the auspices of International Partnerships in Ice Core Sciences (IPICS), defined a priority project to obtain the oldest possible ice core record from Antarctica, an ice core record reaching back to or towards 1.5 million years ago. The deep marine record, the source of most isotopic data, exists only on oceanic plates, which are eventually subducted: the oldest remaining material is 200 million years old. Older sediments are also more prone to corruption by diagenesis. Resolution and confidence in the data decrease over time.

Notable Climate Events in Earth History

Knowledge of precise climatic events decreases as the record goes back in time, but some notable climate events are known:

- Faint young Sun paradox (start).
- Huronian glaciation (~2400 Mya Earth completely covered in ice probably due to Great Oxygenation Event).
- Later Neoproterozoic Snowball Earth (~600 Mya, precursor to the Cambrian Explosion).
- Andean-Saharan glaciation (~450 Mya).
- Carboniferous Rainforest Collapse (~300 Mya).
- Permian–Triassic extinction event (251.4 Mya).
- Oceanic Anoxic Events (~120 Mya, 93 Mya, and others).
- Cretaceous–Paleogene extinction event (66 Mya).
- Paleocene–Eocene Thermal Maximum (Paleocene–Eocene, 55Mya).
- Younger Dryas/The Big Freeze (~11,000 BC).
- Holocene climatic optimum (~7000–3000 BC).
- Climate changes of 535-536 (535–536 AD).
- Medieval warm period (900–1300).

- Little Ice Age (1300–1800).
- Year Without a Summer (1816).

History of the Atmosphere

Earliest Atmosphere

The first atmosphere would have consisted of gases in the solar nebula, primarily hydrogen. In addition, there would probably have been simple hydrides such as those now found in gas giants like Jupiter and Saturn, notably water vapor, methane and ammonia. As the solar nebula dissipated, the gases would have escaped, partly driven off by the solar wind.

Second Atmosphere

The next atmosphere, consisting largely of nitrogen and carbon dioxide and inert gases, was produced by outgassing from volcanism, supplemented by gases produced during the late heavy bombardment of Earth by huge asteroids. A major part of carbon dioxide emissions were soon dissolved in water and built up carbonate sediments.

Water-related sediments have been found dating from as early as 3.8 billion years ago. About 3.4 billion years ago, nitrogen was the major part of the then stable "second atmosphere". An influence of life has to be taken into account rather soon in the history of the atmosphere because hints of early life forms have been dated to as early as 3.5 billion years ago. The fact that it is not perfectly in line with the 30% lower solar radiance (compared to today) of the early Sun has been described as the "faint young Sun paradox".

The geological record, however, shows a continually relatively warm surface during the complete early temperature record of Earth with the exception of one cold glacial phase about 2.4 billion years ago. In the late Archaean eon, an oxygen-containing atmosphere began to develop, apparent-ly from photosynthesizing cyanobacteria which have been found as stromatolite fossils from 2.7 billion years ago. The early basic carbon isotopy (isotope ratio propor-tions) was very much in line with what is found today, suggesting that the fundamental features of the carbon cycle were established as early as 4 billion years ago.

Third Atmosphere

The constant rearrangement of continents by plate tectonics influences the long-term evolution of the atmosphere by transferring carbon dioxide to and from large continental carbonate stores. Free oxygen did not exist in the atmosphere until about 2.4 billion years ago, during the Great Oxygenation Event, and its appearance is indicated by the end of the banded iron formations. Until then, any oxygen produced by photosynthesis was consumed by oxidation of reduced materials, notably iron. Molecules of free oxygen did not start to accumulate in the atmosphere until the rate of production of oxygen began to exceed the availability of reducing materials. That point was a shift from a reducing atmosphere to an oxidizing atmosphere. O_2 showed major variations until reaching a steady state of more than 15% by the end of the Precambrian. The following time span was the Phanerozoic eon, during which oxygen-breathing metazoan life forms began to appear.

The amount of oxygen in the atmosphere has fluctuated over the last 600 million years, reaching a peak of 35% during the Carboniferous period, significantly higher than today's 21%. Two main processes govern changes in the atmosphere: plants use carbon dioxide from the atmosphere, releasing oxygen and the breakdown of pyrite and volcanic eruptions release sulfur into the atmosphere, which oxidizes and hence reduces the amount of oxygen in the atmosphere. However, volcanic eruptions also release carbon dioxide, which plants can convert to oxygen. The exact cause of the variation of the amount of oxygen in the atmosphere is not known. Periods with much oxygen in the atmosphere are associated with rapid development of animals. Today's atmosphere contains 21% oxygen, which is high enough for rapid development of animals.

Currently, anthropogenic greenhouse gases are accumulating in the atmosphere, which is the main cause of global warming.

Climate During Geological Ages

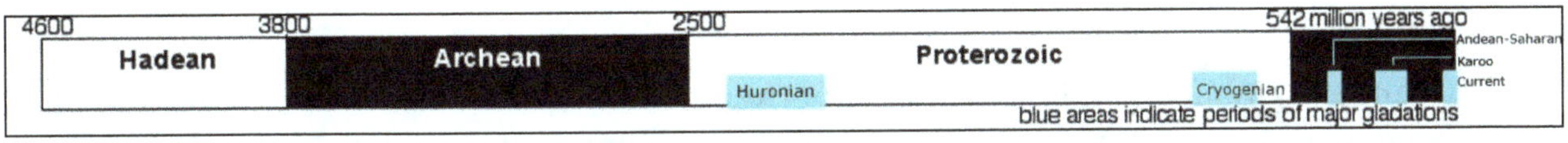

Timeline of glaciations, shown in blue:

- The Huronian glaciation, is the first known glaciation in Earth's history, and lasted from 2400-2100 million years ago.
- The Cryogenian glaciation lasted from 850-635 million years ago.
- The Andean-Saharan glaciation lasted from 450–420 million years ago.
- The Karoo glaciation lasted from 360–260 million years ago.
- The Quaternary glaciation is the current glaciation period and begun 2.58 million years ago.

Precambrian Climate

The climate of the late Precambrian showed some major glaciation events spreading over much of the earth. At this time the continents were bunched up in the Rodinia supercontinent. Massive deposits of tillites are found and anomalous isotopic signatures are found, which gave rise to the Snowball Earth hypothesis. As the Proterozoic Eon drew to a close, the Earth started to warm up. By the dawn of the Cambrian and the Phanerozoic, life forms were abundant in the Cambrian explosion with average global temperatures of about 22 °C.

Phanerozoic Climate

Major drivers for the preindustrial ages have been variations of the sun, volcanic ashes and exhalations, relative movements of the earth towards the sun and tectonically induced effects as for major sea currents, watersheds and ocean oscillations. In the early Phanerozoic, increased atmospheric carbon dioxide concentrations have been linked to driving or amplifying increased global temperatures. Royer et al. 2004 found a climate sensitivity for the rest of the Phanerozoic which was calculated to be similar to today's modern range of values.

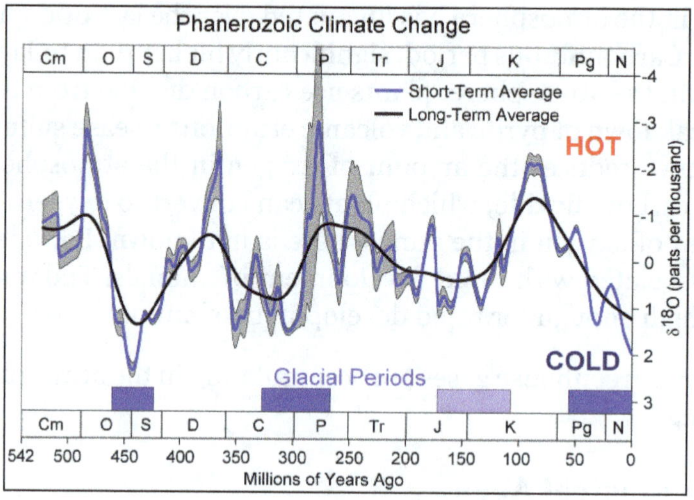

500 million years of climate change.

The difference in global mean temperatures between a fully glacial Earth and an ice free Earth is estimated at approximately 10 °C, though far larger changes would be observed at high latitudes and smaller ones at low latitudes. One requirement for the development of large scale ice sheets seems to be the arrangement of continental land masses at or near the poles. The constant rearrangement of continents by plate tectonics can also shape long-term climate evolution. However, the presence or absence of land masses at the poles is not sufficient to guarantee glaciations or exclude polar ice caps. Evidence exists of past warm periods in Earth's climate when polar land masses similar to Antarctica were home to deciduous forests rather than ice sheets.

The relatively warm local minimum between Jurassic and Cretaceous goes along with an increase of subduction and mid-ocean ridge volcanism due to the breakup of the Pangea supercontinent.

Superimposed on the long-term evolution between hot and cold climates have been many short-term fluctuations in climate similar to, and sometimes more severe than, the varying glacial and interglacial states of the present ice age. Some of the most severe fluctuations, such as the Paleocene-Eocene Thermal Maximum, may be related to rapid climate changes due to sudden collapses of natural methane clathrate reservoirs in the oceans.

A similar, single event of induced severe climate change after a meteorite impact has been proposed as reason for the Cretaceous–Paleogene extinction event. Other major thresholds are the Permian-Triassic, and Ordovician-Silurian extinction events with various reasons suggested.

Quaternary Climate

The Quaternary sub-era includes the current climate. There has been a cycle of ice ages for the past 2.2–2.1 million years (starting before the Quaternary in the late Neogene Period).

Note in the graphic on the right the strong 120,000-year periodicity of the cycles, and the striking asymmetry of the curves. This asymmetry is believed to result from complex interactions of feedback mechanisms. It has been observed that ice ages deepen by progressive steps, but the recovery to interglacial conditions occurs in one big step.

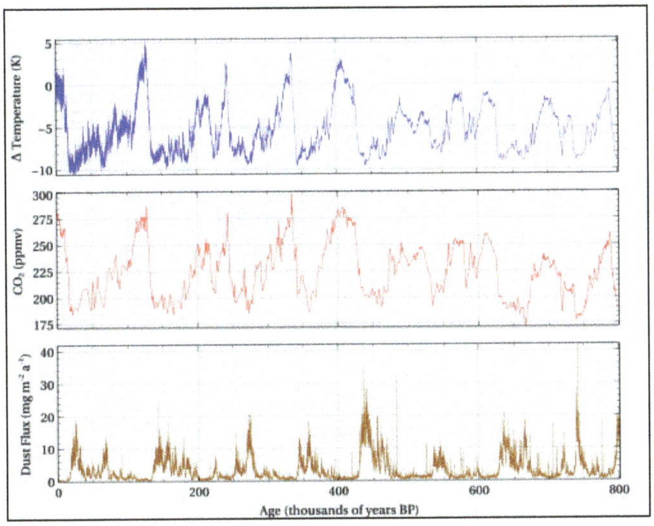

Ice core data for the past 800,000 years . Note length of glacial-interglacial cycles averages ~100,000 years. Blue curve is temperature, red curve is atmospheric CO_2 concentrations, and brown curve is dust fluxes. Today's date is on the left side of the graph because the x-axis values represent "age before 1950".

The graph below shows the temperature change over the past 12 000 years, from various sources. The thick black curve is an average.

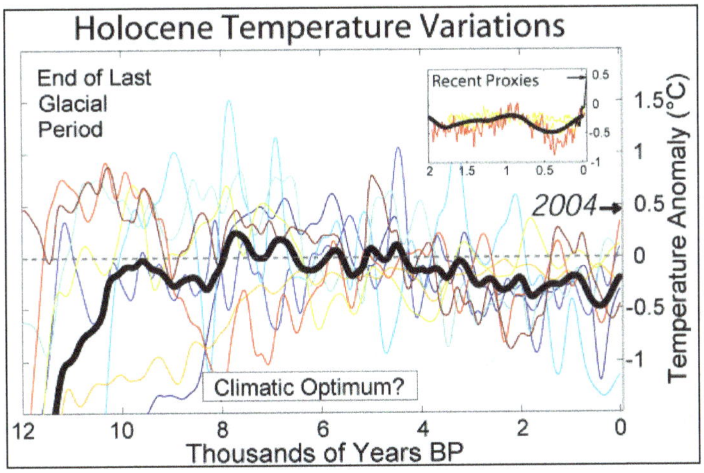

Holocene Temperature Variations.

Climate Forcings

The climate forcing is the difference of radiant energy (sunlight) received by the Earth and the outgoing longwave radiation back to space. The radiative forcing is quantified based on the CO_2 amount in the tropopause, in units of watts per square meter to the Earth's surface. Dependent on the radiative balance of incoming and outgoing energy, the Earth either warms up or cools down. Earth radiative balance originates from changes in solar insolation and the concentrations of greenhouse gases and aerosols. Climate change may be due to internal processes in Earth sphere's and/or following external forcings.

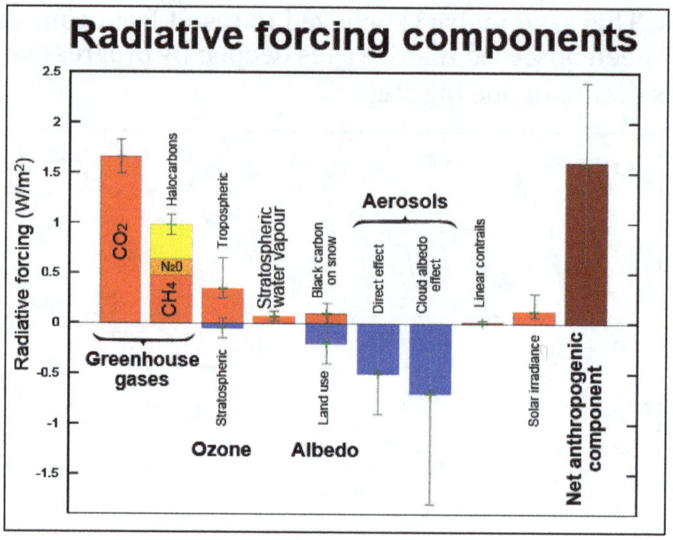

Radiative forcings, IPCC (2007).

Internal Processes and Forcings

The Earth's climate system involves the study of the atmosphere, biosphere, cryosphere, hydrosphere, and lithosphere, and the sum of these processes from Earth sphere's is considered the processes affecting the climate. Greenhouse gases act as the internal forcing of the climate system. Particular interests in climate science and paleoclimatology focuses on the study of Earth climate sensitivity, in response to the sum of forcings.

Examples:

- Thermohaline circulation (Hydrosphere).
- Life (Biosphere).

External Forcings

- The Milankovitch cycles determine Earth distance and position to the Sun. The solar insolation, is the total amount of solar radiation received by Earth.
- Volcanic eruptions, are considered an external forcing.
- Human changes of the composition of the atmosphere or land use.

Mechanisms

On timescales of millions of years, the uplift of mountain ranges and subsequently weathering processes of rocks and soils and the subduction of tectonic plates, are an important part of the carbon cycle. The weathering sequesters CO_2, by the reaction of minerals with chemicals (especially silicate weathering with CO_2) and thereby removing CO_2 from the atmosphere and reducing the radiative forcing. The opposite effect is volcanism, responsible for the natural greenhouse effect, by emitting CO_2 into the atmosphere, thus affecting glaciation (Ice Age) cycles.

James Hansen suggested that humans emit CO_2 10,000 times faster than natural processes have done in the past.

Ice sheet dynamics and continental positions (and linked vegetation changes) have been important factors in the long term evolution of the earth's climate. There is also a close correlation between CO_2 and temperature, where CO_2 has a strong control over global temperatures in Earth history.

Paleotempestology

Paleotempestology is the study of past tropical cyclone activity by means of geological proxies as well as historical documentary records. The term was coined by Kerry Emanuel.

Methods

Sedimentary Proxy Records

Examples of proxies include overwash deposits preserved in the sediments of coastal lakes and marshes, microfossils such as foraminifera, pollen, diatoms, dinoflagellates, phytoliths contained in coastal sediments, wave-generated or flood-generated sedimentary structures or deposits (called tempestites) in marine or lagoonal sediments, storm wave deposited coral shingle, shell, sand and shell and pure sand shore parallel ridges.

The method of using overwash deposits preserved in coastal lake and marsh sediments is adopted from earlier studies of paleotsunami deposits. Both storms and tsunamis leave very similar if not identical sedimentary deposits in coastal lakes and marshes and differentiating between the two in a sedimentary record can be difficult. The first studies to examine prehistoric records of tropical cyclones occurred in Australia and the South Pacific during the late 1970s and early 1980s. These studies examined multiple shore parallel ridges of coral shingle or sand and marine shells. As many as 50 ridges can be deposited at a site with each representing a past severe tropical cyclone over the previous 6,000 years. Tsunamis are not known to deposit multiple sedimentary ridges and therefore these features can be more easily attributed to a past storm at any given site.

Coastal sedimentary analyses have been done at the U.S. Gulf coast, the Atlantic coast from South Carolina up to New Jersey and New England, and the Caribbean Sea. Also, studies on pre-historic tropical cyclones hitting Australia have been made. A study covering the South China Sea coast has also been published.

Markers in Coral

Rocks contain certain isotopes of elements, known as natural tracers, which describe the conditions under which they formed. By studying the calcium carbonate in coral rock, past sea surface temperature and hurricane information can be revealed. Heavier oxygen isotopes (^{18}O) decrease in relation to lighter oxygen isotopes (^{16}O) in coral during periods of very heavy rainfall. Since hurricanes are the main source of extreme rainfall in the tropical oceans, past hurricane events can be dated to the days of their impact on the coral by looking at the decreased ^{18}O concentration within the coral.

Speleothems and Tree Rings

Isotope studies in speleothems and tree rings (dendrochronology) offers a means by which higher resolution records of long-term tropical cyclone histories can be attained. Unlike the isotope records, the sedimentary records are too coarse in their resolution to register quasi-cyclic activity at decadal to centennial scales. These higher resolution records therefore offer a means for possibly differentiating between the natural variability of tropical cyclone behaviour and the effects of anthropogenically induced global climate change. Recent studies with stalagmites in Belize shows that events can be determined on a week-by-week basis.

Historical Records

The *Royal Charter* which sank in a storm off Anglesey in 1859.

Before the invention of the telegraph in the early to mid-19th century, news was as fast as the fastest horse or stagecoach or ship. Normally, there was no advance warning of a tropical cyclone impact. However, the situation changed in the 19th century as seafaring people and land-based researchers, such as Father Viñes in Cuba, came up with systematic methods of reading the sky's appearance or the sea state, which could foretell a tropical cyclone's approach up to a couple days in advance.

One of the best documented storms is the Royal Charter Storm of 1859 which caused over 800 deaths in the UK alone. It led directly to the formation of the Meteorological Office under Robert Fitzroy. However, wind speed could not be measured accurately at the time, methods only becoming available after the Tay Bridge disaster of 1879. One of the better sources for storms in and around Britain is the shipwreck statistics compiled annually by the Board of Trade, but which have yet to be analysed in detail.

Michael Chenoweth used 18th century journals to reconstruct the climate of Jamaica. Together with Dmitry Divine, he also created a 318-year (1690–2007) record of tropical cyclones in the

Lesser Antilles, using newspaper accounts, ships' logbooks, meteorological journals, and other document sources.

In China, the abundance of historical documentary records in the form of *Fang Zhi* (semiofficial local gazettes) offers an extraordinary opportunity for providing a high-resolution historical dataset for the frequency of typhoon strikes.

Historical Climatology

Historical climatology is the study of historical changes in climate and their effect on human history and development. This differs from paleoclimatology which encompasses climate change over the entire history of Earth. The study seeks to define periods in human history where temperature or precipitation varied from what is observed in the present day. The primary sources include written records such as sagas, chronicles, maps and local history literature as well as pictorial representations such as paintings, drawings and even rock art. The archaeological record is equally important in establishing evidence of settlement, water and land usage.

Techniques of Historical Climatology

In literate societies, historians may find written evidence of climatic variations over hundreds or thousands of years, such as phenological records of natural processes, for example viticultural records of grape harvest dates. In preliterate or non-literate societies, researchers must rely on other techniques to find evidence of historical climate differences.

Past population levels and habitable ranges of humans or plants and animals may be used to find evidence of past differences in climate for the region. Palynology, the study of pollens, can show not only the range of plants and to reconstruct possible ecology, but to estimate the amount of precipitation in a given time period, based on the abundance of pollen in that layer of sediment or ice.

Evidence of Climatic Variations

The eruption of the Toba supervolcano, 70,000 to 75,000 years ago reduced the average global temperature by 5 degrees Celsius for several years and may have triggered an ice age. It has been postulated that this created a bottleneck in human evolution. A much smaller but similar effect occurred after the eruption of Krakatoa in 1883, when global temperatures fell for about 5 years in a row.

Before the retreat of glaciers at the start of the Holocene (~9600 BC), ice sheets covered much of the northern latitudes and sea levels were much lower than they are today. The start of our present interglacial period appears to have helped spur the development of human civilization.

Human Record

Evidence of a warm climate in Europe, for example, comes from archaeological studies of settlement and farming in the Early Bronze Age at altitudes now beyond cultivation, such as Dartmoor,

Exmoor, the Lake district and the Pennines in Great Britain. The climate appears to have deteriorated towards the Late Bronze Age however. Settlements and field boundaries have been found at high altitude in these areas, which are now wild and uninhabitable. Grimspound on Dartmoor is well preserved and shows the standing remains of an extensive settlement in a now inhospitable environment.

The 16th-century Skálholt Map of Norse America.

Some parts of the present Saharan desert may have been populated when the climate was cooler and wetter, judging by cave art and other signs of settlement in Prehistoric Central North Africa.

One of Grimspound's hut circles.

The Medieval Warm Period was a time of warm weather between about AD 800–1300, during the European Medieval period. Archaeological evidence supports studies of the Norse sagas which describe the settlement of Greenland in the 9th century AD of land now quite unsuitable for cultivation. For example, excavations at one settlement site have shown the presence of birch trees during the early Viking period. The same period records the discovery of an area called Vinland, probably in North America, which may also have been warmer than at present, judging by the alleged presence of grape vines. The interlude is known as the Medieval Warm Period.

Little Ice Age

Later examples include the Little Ice Age, well documented by paintings, documents (such as diaries) and events such as the River Thames frost fairs held on frozen lakes and rivers in the 17th and 18th centuries. The River Thames was made more narrow and flowed faster after old London Bridge was demolished in 1831, and the river was embanked in stages during the 19th century, both of which made the river less liable to freezing. Among the earliest references to the coming climate change is an entry in the *Anglo-Saxon Chronicle* dated 1046:

> "And in this same year after the 2nd of February came the severe winter with frost and snow, and with all kinds of bad weather, so that there was no man alive who could remember so severe a winter as that, both through mortality of men and disease of cattle; both birds and fishes perished through the great cold and hunger."

The *Chronicle* is the single most important historical source for the period in England between the departure of the Romans and the decades following the Norman Conquest. Much of the information given in the *Chronicle* is not recorded elsewhere.

The Frozen Thames, 1677.

The Little Ice Age brought colder winters to parts of Europe and North America. In the mid-17th century, glaciers in the Swiss Alps advanced, gradually engulfing farms and crushing entire villages. The River Thames and the canals and rivers of the Netherlands often froze over during the winter, and people skated and even held frost fairs on the ice. The first Thames frost fair was in 1607; the last in 1814, although changes to the bridges and the addition of an embankment affected the river flow and depth, diminishing the possibility of freezes. The freeze of the Golden Horn and the southern section of the Bosphorus took place in 1622. In 1658, a Swedish army marched across the Great Belt to Denmark to invade Copenhagen. The Baltic Sea froze over, enabling sledge rides from Poland to Sweden, with seasonal inns built on the way. The winter of 1794/1795 was particularly harsh when the French invasion army under Pichegru could march on the frozen rivers of the Netherlands, while the Dutch fleet was fixed in the ice in Den Helder harbour. In the winter of 1780, New York Harbour froze, allowing people to walk from Manhattan to Staten Island. Sea ice surrounding Iceland extended for miles in every direction, closing that island's harbours to shipping.

The last written records of the Norse Greenlanders are from a 1408 marriage in Hvalsey Church — today the best-preserved of the Norse ruins.

The severe winters affected human life in ways large and small. The population of Iceland fell by half, but this was perhaps also due to fluorosis caused by the eruption of the volcano Laki in 1783. Iceland also suffered failures of cereal crops and people moved away from a grain-based diet. The Norse colonies in Greenland starved and vanished (by the 15th century) as crops failed and livestock could not be maintained through increasingly harsh winters, though Jared Diamond noted that they had exceeded the agricultural carrying capacity before then. In North America, American Indians formed leagues in response to food shortages. In Southern Europe, in Portugal, snow storms were much more frequent while today they are rare. There are reports of heavy snowfalls in the winters of 1665, 1744 and 1886.

In contrast to its uncertain beginning, there is a consensus that the Little Ice Age ended in the mid-19th century.

Evidence of Anthropogenic Climate Change

Through deforestation and agriculture, some scientists have proposed a human component in some historical climatic changes. Human-started fires have been implicated in the transformation of much of Australia from grassland to desert. If true, this would show that non-industrialized societies could have a role in influencing regional climate. Deforestation, desertification and the salinization of soils may have contributed to or caused other climatic changes throughout human history.

Tornado Climatology

Tornadoes have been recorded on all continents except Antarctica and are most common in the middle latitudes where conditions are often favorable for convective storm development. The United States has the most tornadoes of any country, as well as the strongest and most violent tornadoes. A large portion of these tornadoes form in an area of the central United States popularly known as Tornado Alley. Other areas of the world that have frequent tornadoes include significant portions of Europe, South Africa, Bangladesh, parts of Argentina, Uruguay, and southern Brazil, New Zealand, and far eastern Asia.

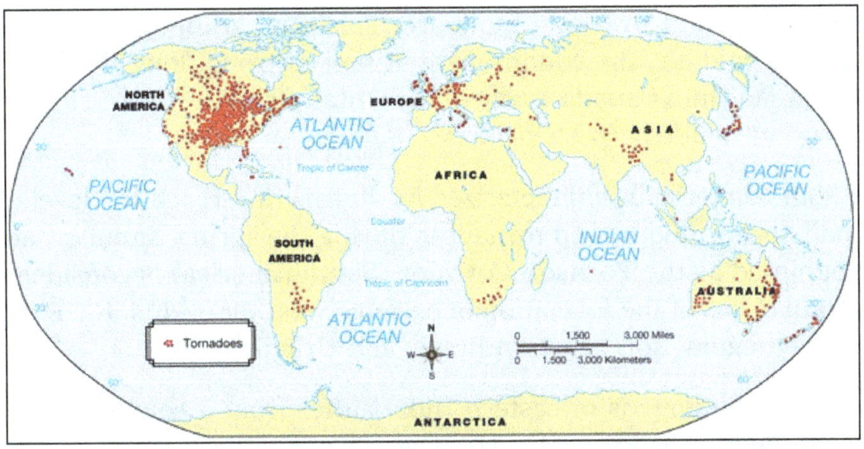

Areas worldwide with the highest frequency of tornadoes are indicated by orange shading.

The United States averaged 1,274 tornadoes per year in the last decade while Canada reports nearly 100 annually (largely in the southern regions). However, the UK has most tornadoes per area per year, 0.14 per 1000 km^2, although these tornadoes are generally weak, and many other European countries have a similar number of tornadoes per area.

The severity of tornadoes is commonly measured by the Enhanced Fujita Scale, which scales tornado intensity from EF0 to EF5 by wind speed and the amount of damage they do to human environments. These judgments are made after the tornado has dissipated and the damage trail is carefully studied by weather professionals.

Tornadoes are most common in spring and least common in winter. The seasonal transition during autumn and spring promotes the development of extratropical cyclones and frontal systems that support strong convective storms. Tornadoes are also common in landfalling tropical cyclones, where they are focused in the right poleward section of the cyclone. Tornadoes can also be spawned as a result of eyewall mesovortices, which persist until landfall. However, favorable conditions for tornado development can occur any time of the year.

Tornado occurrence is highly dependent on the time of day, because of solar heating. Worldwide, most tornadoes occur in the late afternoon, between 3 pm and 7 pm local time, with a peak near 5 pm. Destructive tornadoes can occur at any time of day, as evidenced by the Gainesville Tornado of 1936 (one of the deadliest tornadoes in history) that occurred at 8:30 am local time.

Geography

The United States has the most tornadoes of any country. Many of these form in an area of the central (with some definitions including Southern) United States known as Tornado Alley. This area extends into Canada, particularly the Prairie Provinces and Ontario; however, activity in Canada is less frequent and intense than that of the US. The high frequency of tornadoes in North America is largely due to geography, as moisture from the Gulf of Mexico is easily advected into the mid-continent with few topographic barriers in the way. The Rocky Mountains block Pacific-sourced moisture and buckle the atmospheric flow, forcing drier air at mid-levels of the troposphere due to downsloping winds and causing cyclogenesis downstream to the east of the mountains. Downsloping winds off the Rockies force the formation of a dry line when the flow aloft is strong, while the Gulf of Mexico fuels abundant low-level moisture. This unique topography allows for frequent collisions of warm and cold air, the conditions that breed strong, long-lived storms throughout the year. This area extends into Canada, particularly Ontario and the Prairie Provinces, and strong tornadoes can also occur in northern Mexico.

A large region of South America is characterized by storms that reach the level of supercells and produce intense hailstorms, floods, and tornadoes during the spring, summer, and early fall. The region recently appointed as the Tornado Corridor (South America) is considered as the second largest in the world in terms of the formation of extreme weather events. It covers most of central Argentina, southern Paraguay, southeastern Brazil, and Uruguay.

Bangladesh and surrounding areas of eastern India suffer from a couple tornadoes annually of similar severity to stronger tornadoes in the US. These occur with a greater recurrence interval (although over a smaller region), and tend to be under-reported due to the scarcity of media coverage of a developing country. The annual human death toll from tornadoes in Bangladesh is estimated at about 179 deaths per year, which is much greater than in the US. This is likely due to the density of population, poor quality of construction, lack of tornado safety knowledge and warnings, and other factors.

Other areas of the world that have frequent strong tornadoes include Germany, the Czech Republic, Slovakia, Italy, Spain, China, and the Philippines. Australia, France, Russia, areas of the Middle East, Japan, and parts of Mexico have a history of multiple damaging tornado events.

Tornadoes in the USA

The United States averaged 1,274 tornadoes per year in the last decade. April 2011 saw the most tornadoes ever recorded for any month in the US National Weather Service's history, 875; the previous record was 542 in one month. It has more tornadoes yearly than any other country and reports more violent (F4 and F5) tornadoes than anywhere else.

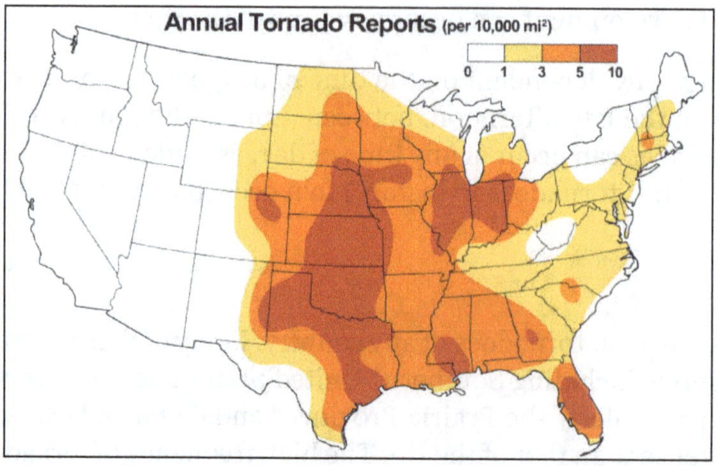

Average annual tornado reports in the United States.

Tornadoes are common in many states but are most common to the west of the Appalachian Mountains and to the east of the Rockies. The Atlantic seaboard states – North Carolina, South Carolina, Georgia and Virginia – are also very vulnerable, as well as Florida. The areas most vulnerable to tornadoes are the Southern Plains and Florida, though most Florida tornadoes are relatively weak. The Southern USA is one of the worst affected regions in terms of casualties.

Tornado reports have been officially collated since 1950. These reports have been gathered by the National Climatic Data Center (NCDC), based in Asheville, North Carolina. A tornado can be reported more than once, such as when a storm crosses a county line and reports are made from two counties.

Common Misconceptions

Some people mistakenly believe that tornadoes only occur in the countryside. This is hardly the case. While it is true that the plains states are the most tornado-prone places in the nation, it should be noted that tornadoes have been reported in every U.S state, including Alaska and Hawaii. One likely reason why tornadoes are so common in the central U.S is because this is where Arctic air first collides with warm tropical air from the Gulf of Mexico where the cold front has not been "weakened" yet. As it heads further east, however, it is possible for the front to lose its strength as it travels over more warm air. Therefore, tornadoes are not as common on the East Coast as they are in the Midwest. However, they have happened on rare occasion, such as the F3 twister that struck Limerick Township, Pennsylvania near Philadelphia on 27 July 1994; the F2 twister that struck the northern suburbs of New York City on 12 July 2006; the EF2 twister in the borough of Brooklyn on 8 August 2007; or the F4 twister that struck La Plata, Maryland on 28 April 2002.

Tornadoes can occur west of the continental divide, but they are infrequent and usually relatively weak and short-lived. Recently tornadoes have struck the Pacific coast town of Lincoln City, Oregon (1996); Sunnyvale, California (1998); and downtown Salt Lake City, Utah (1999). The California Central Valley is an area of some frequency for tornadoes, albeit of weak intensity. Though tornadoes that occur on the Western Seaboard typically are weak, more powerful, damaging tornadoes can occur, such as a tornado that occurred on 22 May 2008 in Per-ris, California.

More tornadoes occur in Texas than in any other US state. The state which has the highest number of tornadoes per unit area is Florida, although most of the tornadoes in Florida are weak tornadoes of EF0 or EF1 intensity. A number of Florida's tornadoes occur along the edge of hurricanes that strike the state. The state with the highest number of strong tornadoes per unit area is Oklahoma. The neighboring state of Kansas is also a particularly notorious tornado state. It records the most EF4 and EF5 tornadoes in the country.

Tornadoes in Canada

Canada also experiences numerous tornadoes, although fewer than the United States. In Canada, at least 80–100 tornadoes occur annually (with many more likely undetected in large expanses of unpopulated areas), causing tens of millions of dollars in damage. Most are weak F0 or F1 in intensity, but there are on average a few F2 or stronger that touch down each season.

For example, the tornado frequency of Southwestern Ontario is about half that of the most prone areas of the central US plains. The last multiple tornado-related deaths in Canada were caused by a tornado in Ear Falls, Ontario, on 9 July 2009, where 3 died, and the last killer tornado was on 21 August 2011 in Goderich, Ontario. The two deadliest tornadoes on Canadian soil were the Regina Cyclone of 30 June 1912 (28 fatalities) and the Edmonton Tornado of 31 July 1987 (27 fatalities). Both of these storms were rated an F4 on the Fujita scale. The city of Windsor was struck by strong tornadoes four times within a 61-year span (1946, 1953, 1974, 1997) ranging in strength from an F2 to F4. Windsor has been struck by more significant tornadoes than any other city in Canada. Canada's first official F5 tornado struck Elie, Manitoba on 22 June 2007. At least two other tornadoes in Saskatchewan in earlier parts of the 20th century are suspected as F5. Tornadoes are most frequent in the provinces of Alberta, Saskatchewan, Manitoba, and Ontario.

Europe

Europe has about 300 tornadoes per year – much more than estimated by Alfred Wegener in his classic book *Wind- und Wasserhosen in Europa* ("Tornadoes and Waterspouts in Europe"). They are most common in June–August, especially in the inlands – rarest in January–March. Strong and violent tornadoes (F3–F5) do occur, especially in some of the interior areas and in the south – but are not as common as in parts of the USA. As in the USA, tornadoes are far from evenly distributed. Europe has some small "tornado alleys" – probably because of frontal collisions as in the south and east of England, but also because Europe is partitioned by mountain ranges like the Alps. Parts of Styria (Steiermark) in Austria may be such a tornado alley, and this county has had at least three F3 tornadoes since 1900. F3 and perhaps one F4 tornado have occurred as far north as Finland.

Since 1900, deadly tornadoes have occurred in Austria, Belgium, Cyprus, England, Finland, France, Germany, Italy (such as the F5/T10 of Udine-Treviso on 24 July 1930, which killed 23 people,), Malta, Poland, Portugal (such as the F3/T7 of Castelo Branco on 6 November 1954, which killed 5 and injured 220), Romania, Russia, Scotland, and Wales. The 1984 Ivanovo–Yaroslavl outbreak, with more than 400 fatalities and 213 injured, was the century's deadliest tornado or outbreak in Europe. It included at least one F5 and one F4. Europe's perhaps deadliest tornado ever (and probably one of the World's deadliest tornadoes) hit Malta in 1551 (or 1556) and killed about 600.

One notable tornado of recent years was the Birmingham Tornado (UK) which struck Birmingham, United Kingdom, in July 2005. A row of houses was destroyed, but no one was killed. A strong F3 (T7) Tornado hit the small town Micheln in Saxony-Anhalt, Germany on 23 July 2004 leaving 6 people injured and more than 250 buildings massively damaged.

Asia

Bangladesh and the eastern parts of India are very exposed to destructive tornadoes causing higher deaths and injuries. Bangladesh, Philippines, and Japan has the highest number of reported tornadoes in Asia.

In June of 2016, a storm producing multiple tornadoes and hail struck a densely populated area of farms and factories near the city of Yancheng in Jiangsu province, about 800 kilometers (500 miles) south of Beijing, killing at least 78 people and destroying buildings. Nearly 500 people were injured, 200 of them critically, the official Xinhua News Agency reported.

South America

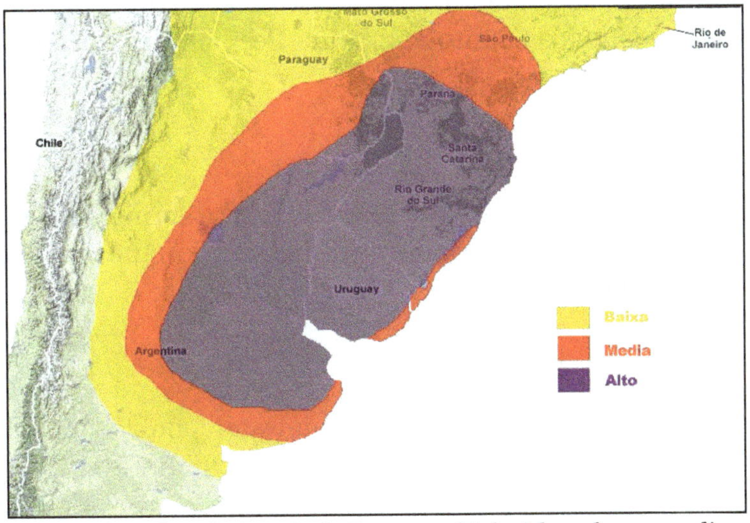

Zone of tornadoes (South America). Gray zone high risk, red zone medium risk.

South America has the hall of tornadoes composed by central and northern Argentina, southern Brazil, Uruguay, and part of Paraguay, considered the second highest frequency tornado region in the world. Argentina has pockets with high tornadic activity, and has the strongest tornadoes in the southern hemisphere like the F5 registered in San Justo, and the tornado outbreak in Buenos Aires with more than 300 tornadoes registered in less than 24 hours. Argentina has the second most tornadoes on the American continent and the world (after the United States) with approximately 300 tornadoes every year depending on the El Niño current, the La Niña current, or neutral year . This region is propitious for tornadoes and severe thunderstorms, due to the large size of the Pampas Plain where the cold air from Patagonia and Antarctica, collides with warm, moist air from areas of Brazil, region northern Argentina and Paraguay, this is added to the dry air from the Andes which collide in this region producing intense storms that frequently producing tornadic supercells. A study of NASA found to the strongest storm occur in the east of the Andes Mountains in Argentina.

South American Tornados

On 16 September 1816, one of the first tornados recorded destroyed the town of Rojas (240 km west of the city of Buenos Aires) On 20 September 1926, an EF4 tornado struck the city of Encarnación (Paraguay) and killed over 300 people, making it the second deadliest tornado in South America. On 21 April 1970, an F4 struck Fray Marcos in Uruguay and killed 11, making it the strongest in Uruguay's history. On 10 January 1973, an F5 struck the city of San Justo, Argentina, 105 km north of the city of Santa Fe. The San Justo tornado is considered the worst tornado ever to occur in the Southern Hemisphere, with winds that exceeded 400 km/h. On 6 May 1992, the people of López Station, Argentina were devastated by an EF4 tornado that caused 4 deaths among the 150 residents. On 13 April 1993, the province of Buenos Aires (Argentina) was impacted by the largest tornado outbreak in South American history. More than 300 tornadoes were recorded over 24 hours with intensities ranging from F1 to F3. On 28 October 1978, an EF4 tornado with winds of 270 km/h hit the city of Mortars in the province of Córdoba, killing 5 people. On 26 December 2003 the 2003 Córdoba Tornado struck 6 km west of Córdoba's city center. It was rated an F3 with winds exceeding 300 km/h, killing 5 and injuring hundreds. On the night of 7 September 2009, an EF4 tornado destroyed the town of San Pedro, Argentina (Misiones), killing 11 people. The same tornado hit the nearby town of Guaraciaba, Brazil, killing 6. The neighboring towns Veloso Santo and Santa Cecilia were seriously damaged and were declared in a state of emergency.

Africa

Tornadoes do occur in South Africa. In October 2011 (i.e. in the spring), two people were killed and nearly 200 were injured after a tornado formed, near Ficksburg in the Free State; more than 1,000 shacks and houses were flattened.

Oceania

Australia has about 16 tornadoes per year – excluding waterspouts, which are common. In New Zealand, a tornado hit the northern suburbs of Auckland on 3 May 2011, killing one and injuring at least 16 people.

Frequency of Occurrence

Tornadoes can form in any month, providing the conditions are favorable. For example, a freak tornado hit South St. Louis County Missouri on 31 December 2010, causing pockets of heavy damage to a modest area before dissipating. The temperature was unseasonably warm that day. They are least common during the winter and most common in spring. Since autumn and spring are transitional periods (warm to cool and vice versa) there are more chances of cooler air meeting with warmer air, resulting in thunderstorms. Tornadoes in the late summer and fall can also be caused by hurricane landfall.

Not every thunderstorm, supercell, squall line, or tropical cyclone will produce a tornado. Precisely the right atmospheric conditions are required for the formation of even a weak tornado. On the other hand, 700 or more tornadoes a year are reported in the contiguous United States.

On average, the United States experiences 100,000 thunderstorms each year, resulting in more than 1,200 tornadoes and approximately 50 deaths per year. The deadliest U.S. tornado recorded is the 18 March 1925, Tri-State Tornado that swept across southeastern Missouri, southern Illinois and southern Indiana, killing 695 people. The biggest tornado outbreak on record—with 353 tornadoes over the course of just 3 1/2 days, including four F5 and eleven F4 tornadoes—occurred starting on 25 April 2011 and intensifying on 26 April and especially the record-breaking day of 27 April before ending on 28 April. It is referred to as the 2011 Super Outbreak. Previously, the record was 148 tornadoes, dubbed the 1974 Super Outbreak. Another such significant storm system was the Palm Sunday tornado outbreak of 1965, which affected the United States Midwest on 11 April 1965. A series of continuous tornado outbreaks is known as a tornado outbreak sequence, with significant occurrences in May 1917, 1930, 1949, and 2003.

Time of Occurrence

Diurnality

Tornado occurrence is highly dependent on the time of day. Austria, Finland, Germany, and the

United States' peak hour of occurrence is 5 pm, with roughly half of all tornado occurrence between 3 p.m. and 7 p.m. local time, due to this being the time of peak atmospheric heating, and thus the maximum available energy for storms; some researchers, including Howard B. Bluestein of the University of Oklahoma, have referred to this phenomenon as "five o'clock magic." Despite this, there are several morning tornadoes reported, like the Seymour, Texas one in April 1980. The most recent happened on 25 May 2015 an F3 tornado raged through Acuna City a Mexico border city at 5:45 am. Killing 14 people.

Seasonality

The time of year is a big factor of the intensity and frequency of tornadoes. On average, in the United States as a whole, the month with the most tornadoes is May, followed by the months June, April, and July. There is no "tornado season" though, as tornadoes, including violent tornadoes and major outbreaks, can and do occur anywhere at any time of year if favorable conditions develop. Major tornado outbreaks have occurred in every month of the year.

July is the peak month in Austria, Finland, and Germany. On average, there are around 294 tornadoes throughout the United States during the month of May, and as many as 543 tornadoes have been reported in the month of May alone (in 2003). The months with the fewest tornadoes are usually December and January, although major tornado outbreaks can and sometimes do occur even in those months. In general, in the Midwestern and Plains states, springtime (especially the month of May) is the most active season for tornadoes, while in the far northern states (like Minnesota and Wisconsin), the peak tornado season is usually in the summer months (June and July). In the colder late autumn and winter months (from early December to late February), tornado activity is generally limited to the southern states, where it is possible for warm Gulf of Mexico air to penetrate.

The reason for the peak period for tornado formation being in the spring has much to do with temperature patterns in the U.S. Tornadoes often form when cool, polar air traveling southeastward from the Rockies overrides warm, moist, unstable Gulf of Mexico air in the eastern states. Tornadoes therefore tend to be commonly found in front of a cold front, along with heavy rains, hail, and damaging winds. Since both warm and cold weather are common during the springtime, the conflict between these two air masses tends to be most common in the spring. As the weather warms across the country, the occurrence of tornadoes spreads northward. Tornadoes are also common in the summer and early fall because they can also be triggered by hurricanes, although the tornadoes caused by hurricanes are often much weaker and harder to spot. Winter is the least common time for tornadoes to occur, since hurricane activity is virtually non-existent at this time, and it is more difficult for warm, moist maritime tropical air to take over the frigid Arctic air from Canada, occurrences are found mostly in the Gulf states and Florida during winter (although there have been some notable exceptions). Interestingly, there is a second active tornado season of the year, late October to mid-November. Autumn, like spring, is a time of the year when warm weather alternates with cold weather frequently, especially in the Midwest, but the season is not as active as it is during the springtime and tornado frequencies are higher along the Atlantic Coastal plain as opposed to the Midwest. They usually appear in late summer.

Long-term Trends

The reliable climatology of tornadoes is limited in geographic and temporal scope; only since

1976 in the United States and 2000 in Europe have thorough and accurate tornado statistics been logged. However, some trends can be noted in tornadoes causing significant damage in the United States, as somewhat reliable statistics on damaging tornadoes exist as far back as 1880. The highest incidence of violent tornadoes seems to shift from the Southeastern United States to the southern Great Plains every few decades. Also, the 1980s seemed to be a period of unusually low tornado activity in the United States, and the number of multi-death tornadoes decreased every decade from the 1920s to the 1980s, suggesting a multi-decadal pattern of some sort.

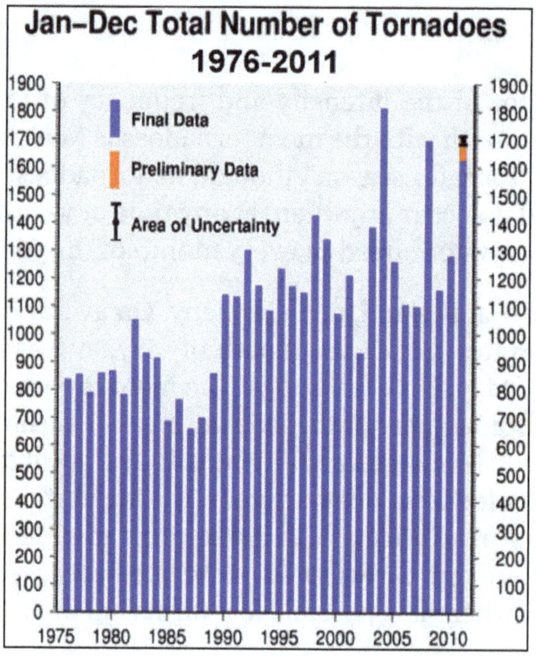

U. S. Annual January – December Tornado Count 1976–2011.

Tropical Cyclone Rainfall Climatology

A tropical cyclone rainfall climatology is developed to determine rainfall characteristics of past tropical cyclones. A tropical cyclone rainfall climatology can be used to help forecast current or upcoming tropical cyclone impacts. The degree of a tropical cyclone rainfall impact depends upon speed of movement, storm size, and degree of vertical wind shear. One of the most significant threats from tropical cyclones is heavy rainfall. Large, slow moving, and non-sheared tropical cyclones produce the heaviest rains. The intensity of a tropical cyclone appears to have little bearing on its potential for rainfall over land, but satellite measurements over the last several years show that more intense tropical cyclones produce noticeably more rainfall over water. Flooding from tropical cyclones remains a significant cause of fatalities, particularly in low-lying areas.

Anticipating a Flood Event

While inland flooding is common to tropical cyclones, there are factors which lead to excessive rainfall from tropical cyclones. Slow motion, as was seen during Hurricane Danny (1997) and Hurricane Wilma, can lead to high amounts of rainfall. The presence of mountains/hills near the coast, like across much of Mexico, Haiti, the Dominican Republic, Central America, Madagascar, Réunion, China, and Japan acts to magnify rainfall potential due to forced upslope flow into the mountains. Strong upper level forcing from a trough moving through the Westerlies and its associated cold front, as was the case during Hurricane Floyd, can lead to high amounts even from systems moving at an average forward motion. Larger tropical cyclones drop more rainfall as they precipitate upon one spot for a longer time frame than average or small tropical cyclones. A combination of two of these factors could be especially crippling, as was seen during Hurricane Mitch in Central America. During the 2005 season, flooding related to slow-moving Hurricane Stan's broad circulation led to 1,662–2,000 deaths.

General Distribution within a Tropical Cyclone

Rainfall Rate per day within radius of the center (Riehl)			
Radius (mi)	Radius (km)	Amount (in)	Amount (mm)
35	56	33.98	863
70	112	13.27	337
140	224	4.25	108
280	448	1.18	30

Isaac Cline was the first to investigate rainfall distribution around tropical cyclones in the early 1900s. He found that a larger proportion of rainfall falls in advance of the center (or eye) than after the center's passage, with the highest percentage falling in the right front quadrant. Father Viñes of Cuba found that some tropical cyclones have their highest rainfall rates in the rear quadrant within a training (non-moving) inflow band. Normally, as a tropical cyclone intensifies, its heavier rainfall rates become more concentrated around its center. Rainfall is found to be heaviest in tropical cyclone's inner core, whether it be the eyewall or central dense overcast, within a degree latitude of the center, with lesser amounts farther away from the center. Most of the rainfall in tropical cyclones is concentrated within its radius of gale-force (34 knots/39 mph/63 km/h) winds. Rainfall is more common near the center of tropical cyclones overnight. Over land, outer bands are more active during the heating of the day, which can act to restrict inflow into the center of the cyclone. Recent studies have shown that half of the rainfall within a tropical cyclone is stratiform in nature. The chart to the right was developed by Riehl in 1954 using meteorological equations that assume a gale radius of about 140 miles (230 km), a fairly symmetric cyclone, and does not consider topographic effects or vertical wind shear. Local amounts can exceed this chart by a factor of two due to topography. Wind shear tends to lessen the amounts below what is shown on the table.

Relation to Storm Size

Larger tropical cyclones have larger rain shields, which can lead to higher rainfall amounts farther from the cyclone's center. This is generally due to the longer time frame rainfall falls at any one spot in a larger system, when compared to a smaller system. Some of the difference seen concerning rainfall between larger and small storms could be the increased sampling of rainfall within a larger tropical cyclone when compared to that of a compact cyclone; in other words, the difference could be the result of a statistical problem.

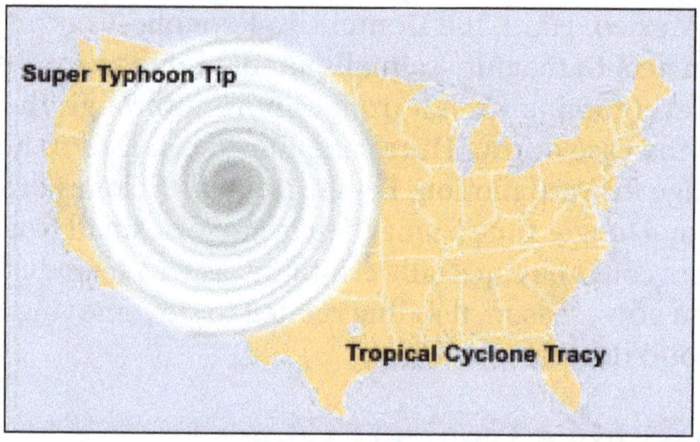

The relative sizes of Typhoon Tip, Cyclone Tracy, and the United States.

Slow/Looping Motion on Rainfall Magnitude

Storms which have moved slowly, or loop, over a succession of days lead to the highest rainfall amounts for several countries. Riehl calculated that 33.97 inches (863 mm) of rainfall per day can be expected within one-half degree, or 35 miles (56 km), of the center of a mature tropical cyclone. Many tropical cyclones progress at a forward motion of 10 knots, which would limit the duration of this excessive rainfall to around one-quarter of a day, which would yield about 8.50 inches (216 mm) of rainfall. This would be true over water, within 100 miles (160 km) of the coastline, and outside topographic features. As a cyclone moves farther inland and is cut off from its supply of warmth and moisture (the ocean), rainfall amounts from tropical cyclones and their remains decrease quickly.

Vertical Wind Shear Impact on Rainfall Shield

Vertical wind shear forces the rainfall pattern around a tropical cyclone to become highly asymmetric, with most of the precipitation falling to the left and downwind of the shear vector, or downshear left. In other words, southwesterly shear forces the bulk of the rainfall north-northeast of the center. If the wind shear is strong enough, the bulk of the rainfall will move away from the center leading to what is known as an exposed circulation center. When this occurs, the potential magnitude of rainfall with the tropical cyclone will be significantly reduced.

Effect of Interaction with Frontal Boundaries/Upper Level Troughs

As a tropical cyclone interacts with an upper-level trough and the related surface front, a distinct northern area of precipitation is seen along the front ahead of the axis of the upper level trough. This type of interaction can lead to the appearance of the heaviest rainfall falling along and to the left of the tropical cyclone track, with the precipitation streaking hundreds of miles or kilometers downwind from the tropical cyclone. The stronger the upper trough picking up the tropical cyclone, the more significant the left of track shift in the rainfall distribution tends to be.

Mountains

Moist air forced up the slopes of coastal hills and mountain chains can lead to much heavier rainfall than in the coastal plain. This heavy rainfall can lead to landslides, which still cause significant loss of life such as seen during Hurricane Mitch in Central America.

Global Distribution

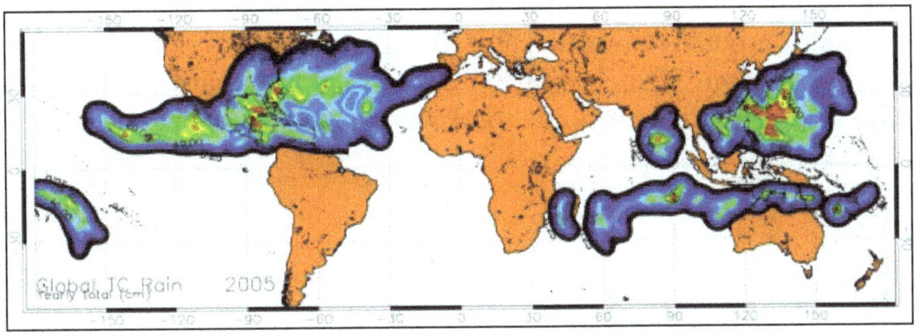

Global tropical cyclone rainfall in 2005.

Globally, tropical cyclone rainfall is more common across the northern hemisphere than across the southern hemisphere. This is mainly due to the normal annual tropical cyclone distribution, as between half and two-thirds of all tropical cyclones form north of the equator. Rainfall is concentrated near the 15th parallel in both hemispheres, with a less steep dropoff seen with latitude across the northern hemisphere, due to the stronger warm water currents seen in that hemisphere which allow tropical cyclones to remain tropical in nature at higher latitudes than south of the equator. In the southern hemisphere, rainfall impacts will be most common between January and March, while north of the equator, tropical cyclone rainfall impacts are more common between June and November. Japan receives over half of its rainfall from typhoons.

United States Tropical Cyclone Rainfall Statistics

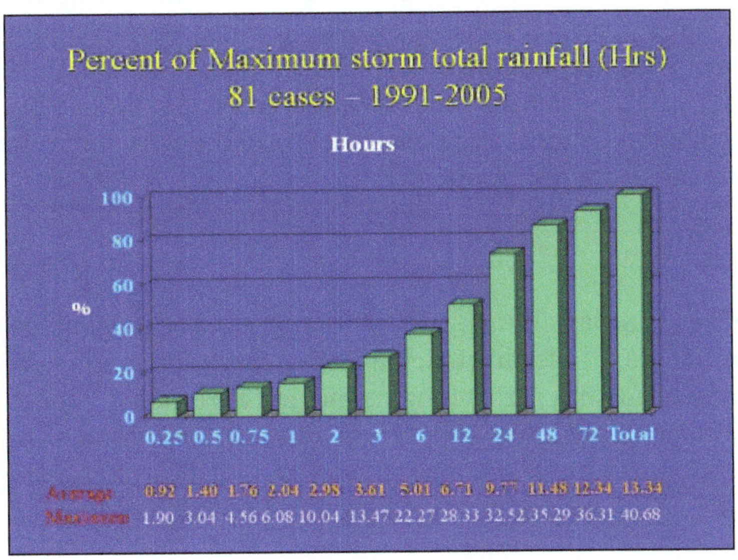

U.S. Tropical Cyclone Rainfall Accumulations per time frame.

Between 1970-2004, inland flooding from tropical cyclones caused a majority of the fatalities in the United States. This statistic changed in 2005, when Hurricane Katrina's impact alone shifted the most deadly aspect of tropical cyclones back to storm surge, which has historically been the most deadly aspect of strong tropical cyclones. On average, five tropical cyclones of at least tropical depression strength lead to rainfall across the contiguous United States annually, contributing around a quarter of the annual rainfall to the southeast United States. While many of these storms form in the Atlantic Basin, some systems or their remnants move through Mexico from the Eastern Pacific Basin. The average storm total rainfall for a tropical cyclone impacting the lower 48 from the Atlantic Basin is about 16 inches (406 mm), with 70–75 percent of the storm total falling within a 24-hour period. The highest point total was seen during Amelia 1978, when 48 inches (1,218 mm) fell upon central Texas.

Urban Climatology

Urban climatology refers to a specific branch of climatology that is concerned with interactions between urban areas and the atmosphere, the effects they have on one another, and the varying spatial and temporal scales at which these processes (and responses) occur.

History

Luke Howard is considered to have established urban climatology with his book *The Climate of London*, which contained continuous daily observations from 1801 to 1841 of wind direction, atmospheric pressure, maximum temperature, and rainfall.

Urban climatology came about as a methodology for studying the results of industrialization and urbanization. Constructing cities changes the physical environment and alters energy, moisture, and motion regimes near the surface. Most of these alterations can be traced to causal factors such as air pollution; anthropogenic sources of heat; surface waterproofing; thermal properties of the surface materials; and morphology of the surface and its specific three-dimensional geometry—building spacing, height, orientation, vegetative layering, and the overall dimensions and geography of these elements. Other factors are relief, proximity to water bodies, size of the city, population density, and land-use distributions.

Influential Factors

Several factors influence the urban climate, including city size, the morphology of the city, land-use configuration, and the geographic setting (such as relief, elevation, and regional climate). Some of the differences between urban and rural climates include air quality, wind patterns, and changes in rainfall patterns, but one of the most studied is the urban heat island effect.

Temperature and Urban Heat Island Effect

Urban environments are typically warmer than their surroundings, as documented over a century ago by Howard. Urban areas are islands or spots on the broader scale compared with more rural surrounding land. The spatial distribution of temperatures occurs in tandem with temporal changes, which are both causally related to anthropogenic sources.

The urban environment has two atmosphere layers, besides the planetary boundary layer outside and extending well above the city: (1) The urban boundary layer is due to the spatially integrated heat and moisture exchanges between the city and its overlying air. (2) The surface of the city corresponds to the level of the urban canopy layer. Fluxes across this plane comprise those from individual units, such as roofs, canyon tops, trees, lawns, and roads, integrated over larger land-use divisions (for example, suburbs). The urban heat island effect has been a major focus of urban climatological studies, and in general the effect the urban environment has on local meteorological conditions.

Pollution

The field also includes the topics of air quality, Radiation Fluxes, Micro-Climates and even issues traditionally associated with architectural design and engineering, such as Wind Engineering. Causes and effects of pollution as understood through Urban Climatology are becoming more important for Urban Planning.

Precipitation

Changes in winds and convection patterns over and around cities impacts precipitation. Contributing factors are believed to be urban heat island, heightened surface roughness, and increased aerosol concentration.

Climate Change

Urban climatology is strongly linked to research surrounding global warming. As centers for socio-economic activities, cities produce large amounts of greenhouse gases, most notably CO_2 as a consequence of human activities such as transport, development, waste related to heating and cooling requirements etc.

Globally, cities are expected to grow into the 21st century (and beyond) - as they grow and develop the landscapes in which they inhabit will change so too will the atmosphere resting above them, increasing emissions of GHG's thus contributing to the global green house effect.

Finally, many cities are vulnerable to the projected consequences of climate change (sea level rise, changes in temperature, precipitation, storm frequency) as most develop on or near coast-lines, nearly all produce distinct urban heat islands and atmospheric pollution: as areas in which there is concentrated human habitation these effects potentially will have the largest and most dramatic impact (e.g. France's heat wave in 2003) and thus are a major focus for urban climatology.

Spatial Planning and Public Health

Urban Climatology impacts decision-making for municipal planning and policy in regards to pollution, extreme heat events, and stormwater modeling.

Dendroclimatology

Dendroclimatology is the science of determining past climates from trees (primarily properties of the annual tree rings). Tree rings are wider when conditions favor growth, narrower when times are difficult. Other properties of the annual rings, such as maximum latewood density (MXD) have been shown to be better proxies than simple ring width. Using tree rings, scientists have estimated many local climates for hundreds to thousands of years previous. By combining multiple tree-ring studies (sometimes with other climate proxy records), scientists have estimated past regional and global climates.

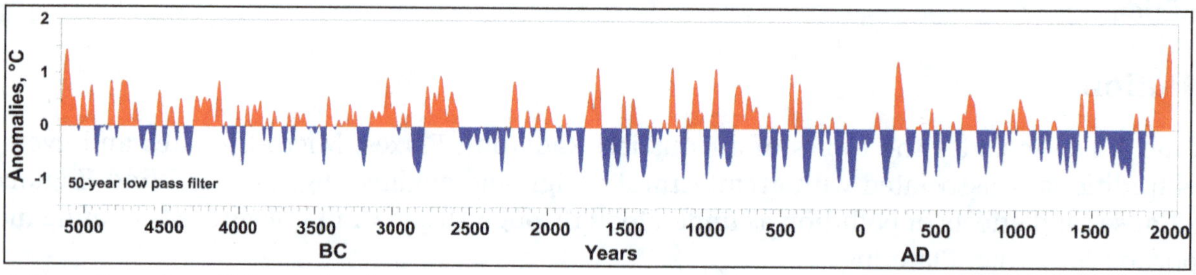

Variation of tree ring width translated into summer temperature anomalies for the past 7000 years, based on samples from holocene deposits on Yamal Peninsula and Siberian now living conifers.

Advantages

Tree rings are especially useful as climate proxies in that they can be well-dated (via matching of the rings from sample to sample, i.e. dendrochronology). This allows extension backwards in time using deceased tree samples, even using samples from buildings or from archeological digs. Another advantage of tree rings is that they are clearly demarked in annual increments, as opposed to other proxy methods such as boreholes. Furthermore, tree rings respond to multiple climatic effects (temperature, moisture, cloudiness), so that various aspects of climate (not just temperature) can be studied. However, this can be a double-edged sword as discussed in Climate factors.

Limitations

Along with the advantages of dendroclimatology are some limitations: confounding factors, geographic coverage, annular resolution, and collection difficulties. The field has developed various methods to partially adjust for these challenges.

Confounding Factors

There are multiple climate and non-climate factors as well as nonlinear effects that impact tree ring width. Methods to isolate single factors (of interest) include botanical studies to calibrate growth influences and sampling of "limiting stands" (those expected to respond mostly to the variable of interest).

Climate Factors

Climate factors that affect trees include temperature, precipitation, sunlight, and wind. To differentiate among these factors, scientists collect information from "limiting stands." An example of a limiting stand is the upper elevation treeline: here, trees are expected to be more affected by temperature variation (which is "limited") than precipitation variation (which is in excess). Conversely, lower elevation treelines are expected to be more affected by precipitation changes than temperature variation. This is not a perfect work-around as multiple factors still impact trees even at the “limiting stand,” but it helps. In theory, collection of samples from nearby limiting stands of different types (e.g. upper and lower treelines on the same mountain) should allow mathematical solution for multiple climate factors. However, this method is rarely used.

Non-climate Factors

Non-climate factors include soil, tree age, fire, tree-to-tree competition, genetic differences, logging or other human disturbance, herbivore impact (particularly sheep grazing), pest outbreaks, disease, and CO_2 concentration. For factors which vary randomly over space (tree to tree or stand to stand), the best solution is to collect sufficient data (more samples) to compensate for confounding noise. Tree age is corrected for with various statistical methods: either fitting spline curves to the overall tree record or using similar aged trees for comparison over different periods (regional curve standardization). Careful examination and site selection helps to limit some confounding effects, for example picking sites undisturbed by modern man.

Non-linear Effects

In general, climatologists assume a linear dependence of ring width on the variable of interest (e.g. moisture). However, if the variable changes enough, response may level off or even turn opposite. The home gardener knows that one can underwater or overwater a house plant. In addition, it is possible that interaction effects may occur (for example “temperature times precipitation” may affect growth as well as temperature and precipitation on their own. Here, also, the “limiting stand” helps somewhat to isolate the variable of interest. For instance, at the upper treeline, where the tree is “cold limited”, it’s unlikely that nonlinear effects of high temperature (“inverted quadratic”) will have numerically significant impact on ring width over the course of a growing season.

Botanical Inferences to Correct for Confounding Factors

Botanical studies can help to estimate the impact of confounding variables and in some cases guide corrections for them. These experiments may be either ones where growth variables are all controlled (e.g. in a greenhouse), partially controlled (e.g. FACE [Free Airborne Concentration Enhancement] experiments—add ref), or where conditions in nature are monitored. In any case, the important thing is that multiple growth factors are carefully recorded to determine what impacts growth. (Insert Fennoscandanavia paper reference). With this information, ring width response can be more accurately understood and inferences from historic (unmonitored) tree rings become more certain. In concept, this is like the limiting stand principle, but it is more quantitative—like a calibration.

Divergence Problem

The divergence problem is the disagreement between the temperatures measured by the thermometers (instrumental temperatures) on one side, and the temperatures reconstructed from the latewood density or width of tree rings on the other side, at many treeline sites in northern forests.

While the thermometer records indicate a substantial warming trend, tree rings from these particular sites do not display a corresponding change in their maximum latewood density or, in some cases, their width. This does not apply to all such studies. Where this applies, a temperature trend extracted from tree rings alone would not show any substantial warming. The temperature graphs calculated from instrumental temperatures and from these tree ring proxies thus "diverge" from one another since the 1950s, which is the origin of the term. This divergence raises obvious questions of whether other, unrecognized divergences have occurred in the past, prior to the era of thermometers. There is evidence suggesting that the divergence is caused by human activities, and so confined to the recent past, but use of affected proxies can lead to overestimation of past temperatures, understating the current warming trend. There is continuing research into explanations and ways to avoid this problem with tree ring proxies.

Geographic Coverage

Trees do not cover the Earth. Polar and marine climates cannot be estimated from tree rings. In perhumid tropical regions, Australia and southern Africa, trees generally grow all year round and don't show clear annual rings. In some forest areas, the tree growth is too much influenced by multiple factors (no "limiting stand") to allow clear climate reconstruction. The coverage difficulty is dealt with by acknowledging it and by using other proxies (e.g. ice cores, corals) in difficult areas. In some cases it can be shown that the parameter of interest (temperature, precipitation, etc.) varies similarly from area to area, for example by looking at patterns in the instrumental record. Then one is justified in extending the dendroclimatology inferences to areas where no suitable tree ring samples are obtainable.

Annular Resolution

Tree rings show the impact on growth over an entire growing season. Climate changes deep in the dormant season (winter) will not be recorded. In addition, different times of the growing season may be more important than others (i.e. May versus September) for ring width. However, in general the ring width is used to infer the overall climate change during the corresponding year (an approximation). Another problem is "memory" or autocorrelation. A stressed tree may take a year or two to recover from a hard season. This problem can be dealt with by more complex modeling (a "lag" term in the regression) or by reducing the skill estimates of chronologies.

Collection Difficulties

Tree rings must be obtained from nature, frequently from remote regions. This means that special efforts are needed to map sites properly. In addition, samples must be collected in difficult (often sloping terrain) conditions. Generally, tree rings are collected using a hand-held borer device, that requires skill to get a good sample. The best samples come from felling a tree and sectioning it. However, this requires more danger and does damage to the forest. It may not be allowed in certain areas, particularly with the oldest trees in undisturbed sites (which are the most interesting scientifically). As with all experimentalists, dendroclimatologists must, at times, decide to make the best of imperfect data, rather than resample. This tradeoff is made more difficult, because sample collection (in the field) and analysis (in the lab) may be separated significantly in time and space. These collection challenges mean that data gathering is not as simple or cheap as conventional laboratory science. However, they also give the field's practitioners much enjoyment, working out of doors, with hands on trees and tools.

Other Measurements

Initial work focused on measuring the tree ring width—this is simple to measure and can be related to climate parameters. But the annual growth of the tree leaves other traces. In particular *maximum latewood density* (MXD) is another metric used for estimating environmental variables. It is, however, harder to measure. Other properties (e.g. isotope or chemical trace analysis) have also been tried most notably by L. M. Libby in her 1974 paper "Temperature Dependence of Isotope Ratios in Tree Rings". In theory, multiple measurements on the same ring will allow differentiation of confounding factors (e.g. precipitation and temperature). However, most studies are still based on ring widths at limiting stands.

Measuring radiocarbon concentrations in tree rings has proven to be useful in recreating past sunspot activity, with data now extending back over 11,000 years.

References

- Roth, David M. (April 29, 2015). "Tropical Cyclone Point Maxima". Tropical Cyclone Rainfall Data. United States Weather Prediction Center. Retrieved 11, February 2020
- Chenoweth, Michael (2003). The 18th century climate of Jamaica derived from the journals of Thomas Thistlewood, 1750-1786. Transactions of the American Philosophical Society. 93. Philadelphia: American Philosophical Society. ISBN 0-87169-932-X
- "Page 1 1 International Partnerships in Ice Core Sciences (IPICS) The oldest ice core: A 1.5 million year record of climate and greenhouse gases from Antarctica" (PDF). Retrieved 22, March 2020
- "Tornado Alley | National Centers for Environmental Information (NCEI) formerly known as National Climatic Data Center (NCDC)". www.ncdc.noaa.gov. Retrieved 15, August 2020
- Hughes, Malcolm K.; Swetman, Thomas W.; Diaz, Henry, eds. (2010). Dendroclimatology: Progress and Prospects. Springer. ISBN 978-1-4020-4010-8
- Luckman, B.H. (2007). "Dendroclimatology". In Elias, Scott A. Encyclopedia of Quaternary Science 1. Elsevier. pp. 465–475. ISBN 978-0-444-51919-1
- Schweingruber, Fritz Hans; Eidgenössische Forschungsanstalt für Wald, Schnee und Landschaft (1996). "Ch. 19". Tree Rings and Environment Dendroecology. Berne: Paul Haupt. ISBN 978-3-258-05458-2
- "U.S. Tornado Climatology | National Centers for Environmental Information (NCEI) formerly known as National Climatic Data Center (NCDC)". www.ncdc.noaa.gov. Retrieved 01, July 2020
- Beerling, David (2007). The emerald planet: how plants changed Earth's history. Oxford University press. p. 47. ISBN 9780192806024
- Yen, Ben Chie. "Urban Stormwater Modeling and Simulation." Bulletin of the American Meteorological Society Apr. 1995: 564+. Academic OneFile. Retrieved 10, May 2020

Chapter 4

Ozone Layer and its Depletion

The ozone layer is a region located in the stratospheric layer of the Earth's atmosphere that absorbs much of the harmful ultraviolet rays emitted by the sun. It contains a high concentration of ozone gas which is instrumental in absorbing the ultraviolet rays. The release of CFCs and halons into the atmosphere results in chemical reactions which break down ozone molecules. The ozone Layer and its depletion are comprehensively covered in this chapter.

Ozone Layer

Ozone layer is a vital component of our atmosphere. Instratosphere, it acts as a protective layer as it absorbs the harmful ultraviolet radiation and does not allow it to reach the earth's surface. Therefore, the presence of the ozone layer is essential factor in the life environment. Ozone is a very reactive molecule and with the help of a catalyst it is easily reduced to the more stable oxygen. The ozone destroying catalysts are natural as well as manmade. The primary cause of ozone depletion is the trace gases mainly chlorine and bromine, from anthropogenic sources.The depletion of ozone layer was discovered for the first time by Farman in 1985. It was established that, ozone 'hole' has occurred in the ozone layer over Antarctica. In 1987, the Montreal protocol set legally binding controls on the production and consumption of gases associated with ozone depletion. The outcome of this response is decline in the total abundance of CFCs in the atmosphere and trend of gradual healing of ozone layer.

The ozone layer was discovered in 1913 by the French physicists Charles Fabry and Henri Buisson. Measurements of the sun showed that the radiation sent out from its surface and reaching the ground on Earth is usually consistent with the spectrum of a black body with a temperature in the range of 5,500–6,000 K (5,227 to 5,727 °C), except that there was no radiation below a wavelength of about 310 nm at the ultraviolet end of the spectrum. It was deduced that the missing radiation was being absorbed by something in the atmosphere. Eventually the spectrum of the missing radiation was matched to only one known chemical, ozone. Its properties were explored in detail by the British meteorologist G. M. B. Dobson, who developed a simple spectrophotometer (the Dobsonmeter) that could be used to measure stratospheric ozone from the ground. Between 1928 and 1958, Dobson established a worldwide network of ozone monitoring stations, which continue to operate to this day. The "Dobson unit", a convenient measure of the amount of ozone overhead, is named in his honor.

The ozone layer absorbs 97 to 99 percent of the Sun's medium-frequency ultraviolet light (from about 200 nm to 315 nm wavelength), which otherwise would potentially damage exposed life forms near the surface.

In 1976 atmospheric research revealed that the ozone layer was being depleted by chemicals released by industry, mainly chlorofluorocarbons (CFCs). Concerns that increased UV radiation due to ozone depletion threatened life on Earth led to bans on the chemicals, and the latest evidence is that ozone depletion has slowed or stopped. The United Nations General Assembly has designated September 16 as the International Day for the Preservation of the Ozone Layer.

Venus also has a thin ozone layer at an altitude of 100 kilometers from the planet's surface.

Sources

The photochemical mechanisms that give rise to the ozone layer were discovered by the British physicist Sydney Chapman in 1930. Ozone in the Earth's stratosphere is created by ultraviolet light striking ordinary oxygen molecules containing two oxygen atoms (O_2), splitting them into individual oxygen atoms (atomic oxygen); the atomic oxygen then combines with unbroken O_2 to create ozone, O_3. The ozone molecule is unstable (although, in the stratosphere, long-lived) and when ultraviolet light hits ozone it splits into a molecule of O_2 and an individual atom of oxygen, a continuing process called the ozone-oxygen cycle. Chemically, this can be described as:

$$O_2 + h\nu_{uv} \rightarrow 2O$$

$$O + O_2 \leftrightarrow O_3$$

About 90 percent of the ozone in the atmosphere is contained in the stratosphere. Ozone concentrations are greatest between about 20 and 40 kilometres (66,000 and 131,000 ft), where they range from about 2 to 8 parts per million. If all of the ozone were compressed to the pressure of the air at sea level, it would be only 3 millimetres ($\frac{1}{8}$ inch) thick.

Ultraviolet light

Although the concentration of the ozone in the ozone layer is very small, it is vitally important to life because it absorbs biologically harmful ultraviolet (UV) radiation coming from the sun. Extremely short or vacuum UV (10–100 nm) is screened out by nitrogen. UV radiation capable of penetrating nitrogen is divided into three categories, based on its wavelength; these are referred to as UV-A (400–315 nm), UV-B (315–280 nm), and UV-C (280–100 nm).

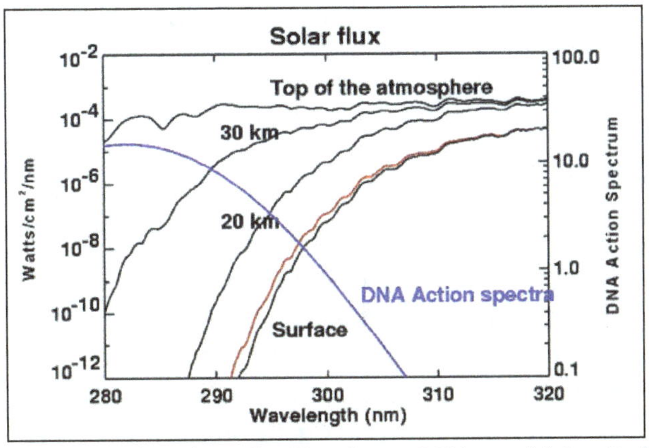

UV-B energy levels at several altitudes. Blue line shows DNA sensitivity. Red line shows surface energy level with 10 percent decrease in ozone.

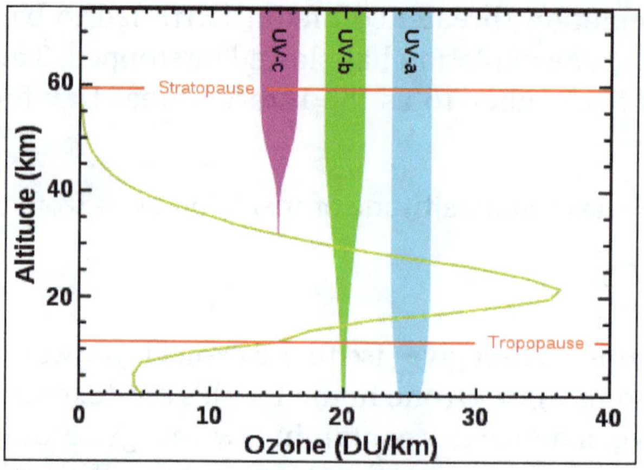

Levels of ozone at various altitudes and blocking of different bands of ultraviolet radiation. Essentially all UVC (100–280 nm) is blocked by dioxygen (from 100–200 nm) or else by ozone (200–280 nm) in the atmosphere.

The shorter portion of the UV-C band and the more energetic UV above this band causes the formation of the ozone layer, when single oxygen atoms produced by UV photolysis of dioxygen (below 240 nm) react with more dioxygen. The ozone layer also blocks most, but not quite all, of the sunburn-producing UV-B (280–315 nm) band, which lies in the wavelengths longer than UV-C. The band of UV closest to visible light, UV-A (315–400 nm), is hardly affected by ozone, and most of it reaches the ground. UV-A does not primarily cause skin reddening, but there is evidence that it causes long-term skin damage.

UV-C, which is very harmful to all living things, is entirely screened out by a combination of dioxygen (< 200 nm) and ozone (> about 200 nm) by around 35 kilometres (115,000 ft) altitude. UV-B radiation can be harmful to the skin and is the main cause of sunburn; excessive exposure can also cause cataracts, immune system suppression, and genetic damage, resulting in problems such as skin cancer. The ozone layer (which absorbs from about 200 nm to 310 nm with a maximal absorption at about 250 nm) is very effective at screening out UV-B; for radiation with a wavelength of 290 nm, the intensity at the top of the atmosphere is 350 million times stronger than at the Earth's surface. Nevertheless, some UV-B, particularly at its longest wavelengths, reaches the surface, and is important for the skin's production of vitamin D.

Ozone is transparent to most UV-A, so most of this longer-wavelength UV radiation reaches the surface, and it constitutes most of the UV reaching the Earth. This type of UV radiation is significantly less harmful to DNA, although it may still potentially cause physical damage, premature aging of the skin, indirect genetic damage, and skin cancer.

Distribution in the Stratosphere

The thickness of the ozone layer—that is, the total amount of ozone in a column overhead—varies by a large factor worldwide, being in general smaller near the equator and larger towards the poles. It also varies with season, being in general thickest during the autumn and thinner during the early spring. The reasons for this latitude and seasonal dependence are complicated, which involve in atmospheric circulation patterns as well as solar intensity. But in general, bitterly cold temperatures mixed with sunlight assist in maximum ozone depletion, which occurs in early spring.

Since stratospheric ozone is produced by solar UV radiation, one might expect to find the highest ozone levels over the tropics and the lowest over polar regions. The same argument would lead one to expect the highest ozone levels in the summer and the lowest in the winter. The observed behavior is very different: most of the ozone is found in the mid-to-high latitudes of the northern and southern hemispheres, and the highest levels are found in the spring, not summer, and the lowest in the autumn, not winter in the northern hemisphere. During winter, the ozone layer actually increases in depth. This puzzle is explained by the prevailing stratospheric wind patterns, known as the Brewer-Dobson circulation. While most of the ozone is indeed created over the tropics, the stratospheric circulation then transports it poleward and downward to the lower stratosphere of the high latitudes. However, owing to the ozone hole phenomenon, the lowest amounts of column ozone found anywhere in the world are over the Antarctic in the southern spring period of September and October and to a lesser extent over the Arctic in the northern spring period of March, April, and May.

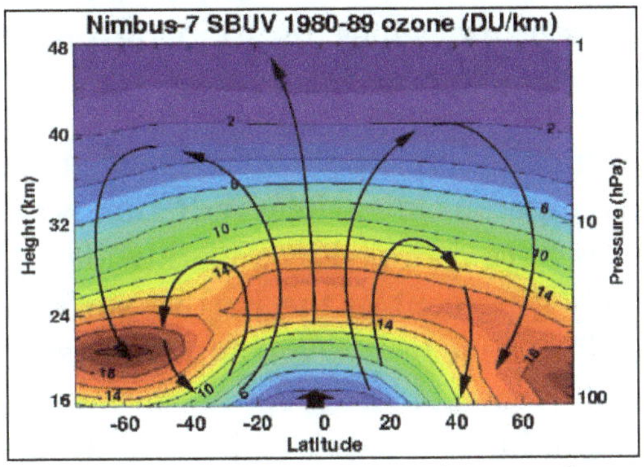

Brewer-Dobson circulation in the ozone layer.

The ozone layer is higher in altitude in the tropics, and lower in altitude outside the tropics, especially in the polar regions. This altitude variation of ozone results from the slow circulation that lifts the ozone-poor air out of the troposphere into the stratosphere. As this air slowly rises in the tropics, ozone is produced as the sun overhead photolyzes oxygen molecules. As this slow circulation levels off and flows towards the mid-latitudes, it carries the ozone-rich air from the tropical middle stratosphere to the lower stratosphere middle and high latitudes . The high ozone concentrations at high latitudes are due to the accumulation of ozone at lower altitudes.

The Brewer-Dobson circulation moves very slowly. The time needed to lift an air parcel by 1 km in the lower tropical stratosphere is about 2 months (18 m per day). However, horizontal poleward transport in the lower stratosphere is much faster and amounts to approximately 100 km per day in the northern hemisphere whilst it is only half as much in the southern hemisphere (~51 km per day). Even though ozone in the lower tropical stratosphere is produced at a very slow rate, the lifting circulation is so slow that ozone can build up to relatively high levels by the time it reaches 26 kilometres (16 mi).

Ozone amounts over the continental United States (25°N to 49°N) are highest in the northern spring (April and May). These ozone amounts fall over the course of the summer to their lowest amounts in October, and then rise again over the course of the winter. Again, wind transport of ozone is principally responsible for the seasonal changes of these higher latitude ozone patterns.

The total column amount of ozone generally increases as we move from the tropics to higher latitudes in both hemispheres. However, the overall column amounts are greater in the northern hemisphere high latitudes than in the southern hemisphere high latitudes. In addition, while the highest amounts of column ozone over the Arctic occur in the northern spring (March–April), the opposite is true over the Antarctic, where the lowest amounts of column ozone occur in the southern spring (September–October).

Depletion

The ozone layer can be depleted by free radical catalysts, including nitric oxide (NO), nitrous oxide (N_2O), hydroxyl (OH), atomic chlorine (Cl), and atomic bromine (Br). While there are natural sources for all of these species, the concentrations of chlorine and bromine increased markedly in recent decades because of the release of large quantities of man-made organohalogen compounds, especially chlorofluorocarbons (CFCs) and bromofluorocarbons. These highly stable compounds are capable of surviving the rise to the stratosphere, where Cl and Br radicals are liberated by the action of ultraviolet light. Each radical is then free to initiate and catalyze a chain reaction capable of breaking down over 100,000 ozone molecules. By 2009, nitrous oxide was the largest ozone-depleting substance (ODS) emitted through human activities.

The breakdown of ozone in the stratosphere results in reduced absorption of ultraviolet radiation. Consequently, unabsorbed and dangerous ultraviolet radiation is able to reach the Earth's surface at a higher intensity. Ozone levels have dropped by a worldwide average of about 4 percent since the late 1970s. For approximately 5 percent of the Earth's surface, around the north and south poles, much larger seasonal declines have been seen, and are described as "ozone holes". The discovery of the annual depletion of ozone above the Antarctic was first announced by Joe Farman, Brian Gardiner and Jonathan Shanklin, in a paper which appeared in *Nature* on May 16, 1985.

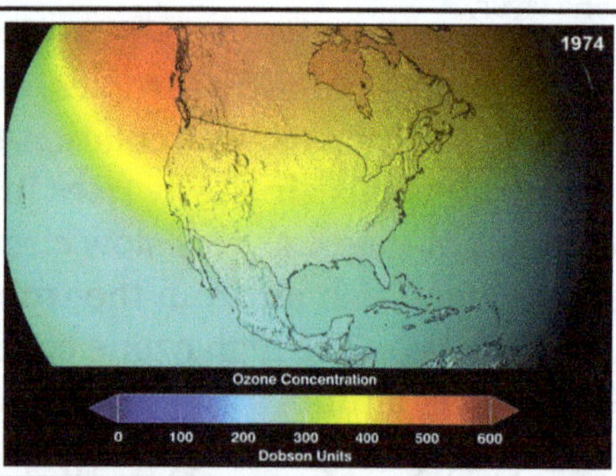

NASA projections of stratospheric ozone concentrations if chlorofluorocarbons had not been banned.

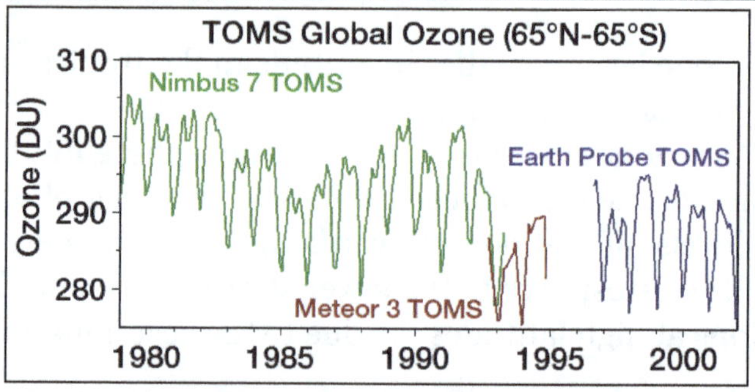

Levels of atmospheric ozone measured by satellite show clear seasonal variations and appear to verify their decline over time.

Regulation

To support successful regulation attempts, the ozone case was communicated to lay persons "with easy-to-understand bridging metaphors derived from the popular culture" and related to "immediate risks with everyday relevance". The specific metaphors used in the discussion (ozone shield, ozone hole) proved quite useful and, compared to global climate change, the ozone case was much more seen as a "hot issue" and imminent risk. Lay people were cautious about a depletion of the ozone layer and the risks of skin cancer.

In 1978, the United States, Canada and Norway enacted bans on CFC-containing aerosol sprays that damage the ozone layer. The European Community rejected an analogous proposal to do the same. In the U.S., chlorofluorocarbons continued to be used in other applications, such as refrigeration and industrial cleaning, until after the discovery of the Antarctic ozone hole in 1985. After negotiation of an international treaty (the Montreal Protocol), CFC production was capped at 1986 levels with commitments to long-term reductions. Since that time, the treaty was amended to ban CFC production after 1995 in the developed countries, and later in developing countries. Today, all of the world's 197 countries have signed the treaty. Beginning January 1, 1996, only recycled and stockpiled CFCs were available for use in developed countries like the US. This production phaseout was possible because of efforts to ensure that there would be substitute chemicals and technologies for all ODS uses.

On August 2, 2003, scientists announced that the global depletion of the ozone layer may be slowing down because of the international regulation of ozone-depleting substances. In a study organized by the American Geophysical Union, three satellites and three ground stations confirmed that the upper-atmosphere ozone-depletion rate slowed down significantly during the previous decade. Some breakdown can be expected to continue because of ODSs used by nations which have not banned them, and because of gases which are already in the stratosphere. Some ODSs, including CFCs, have very long atmospheric lifetimes, ranging from 50 to over 100 years. It has been estimated that the ozone layer will recover to 1980 levels near the middle of the 21st century. A gradual trend toward "healing" was reported in 2016.

Compounds containing C–H bonds (such as hydrochlorofluorocarbons, or HCFCs) have been designed to replace CFCs in certain applications. These replacement compounds are more reactive and less likely to survive long enough in the atmosphere to reach the stratosphere where they could affect the ozone layer. While being less damaging than CFCs, HCFCs can have a negative impact on the ozone layer, so they are also being phased out. These in turn are being replaced by hydrofluorocarbons (HFCs) and other compounds that do not destroy stratospheric ozone at all.

The residual effects of CFCs accumulating within the atmosphere lead to a concentration gradient between the atmosphere and the ocean. This organohalogen compound is able to dissolve into the ocean's surface waters and is able to act as a time-dependent tracer. This tracer helps scientists study ocean circulation by tracing biological, physical and chemical pathways.

Implications for Astronomy

As ozone in the atmosphere prevents most energetic ultraviolet radiation reaching the surface of the Earth, astronomical data in these wavelengths have to be gathered from satellites orbiting above the atmosphere and ozone layer. Most of the light from young hot stars is in the ultraviolet and so study of these wavelengths is important for studying the origins of galaxies. The Galaxy Evolution Explorer, GALEX, is an orbiting ultraviolet space telescope launched on April 28, 2003, which operated until early 2012.

This GALEX image of the Cygnus Loop nebula could not have been taken from the surface of the Earth because the ozone layer blocks the ultra-violet radiation emitted by the nebula.

Ozone Depletion

Ozone is a very reactive molecule and with the help of a catalyst it is easily reduced to the more stable oxygen. The ozone destroying catalysts are natural as well as manmade. Ozone can be destroyed by a number of free radical catalysts such as (i) OH (hydroxyl) -the odd hydrogen atoms and OH come from the dissociation of water vapour, molecular hydrogen and methane (CH_4); (ii) nitric oxide –ozone is destroyed in stratosphere due to the presence nitrogen oxides (NOx, i.e. NO_2 and NO). The source gas of the NOx is nitrous oxide (N_2O) - it is produced by combustion, supersonic jets and fertilizer use; (iii) Chlorine and (iv) Bromine. Anthropogenic activities have sharply increased the levels of chlorine and bromine.

A single chlorine atom has the capability to interact with thousands of ozone molecules in catalytic cycle before it is finally removed. On a per atom basis, the efficiency of bromine is far more than chlorine to destroy ozone, but its abundance is much less in the atmosphere. Therefore, both chlorine and bromine are the major contributors in overall ozone destruction. The increase in abundance of chlorine and bromine during the decades of 1970-90 is reflected in the observed decrease of stratospheric ozone over the Antarctica and other parts. A simplified illustration is given to show ozone destruction below:

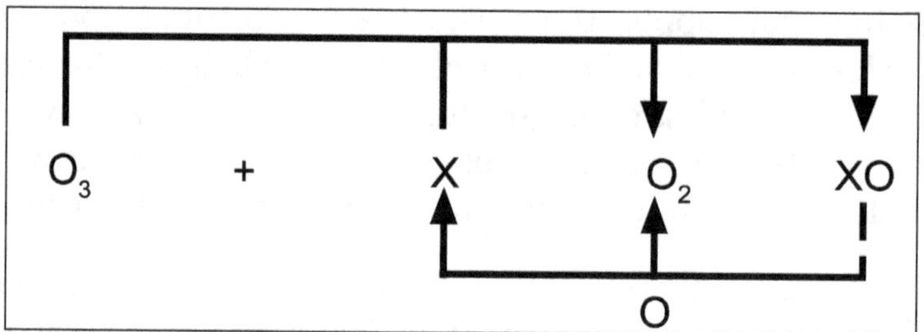

Ozone Destruction Process (where X is any ozone-destroying species, e.g. OH, NO, CR, Br).

The depletion of ozone in the stratosphere due to human activities is a serious global environmental problem. The major offending chemicals are Freons, synthetic compounds containing carbon, fluorine and chlorine atoms. Compounds of this class are also known as *halocarbons*. The popular alternate name for these is*chlorofluorocarbons*, or CFCs. They are versatile compounds that are chemically stable, odorless, nontoxic, noncorrosive, and inexpensive to produce. CFCs, carbon tetrachloride (used in fire extinguishers and solvents), alternates of CFCs - hydrochlorofluorocarbons (HCFCs) and methyl chloroform are the chief chlorine containing gases. These gases are used in many applications such as – refrigeration, air conditioning, foam blowing, aerosol propellants and cleaning of metals and electronic components. Methyl chloroform is used in industries for cold cleaning, vapor degreasing, chemical processing, adhesives and aerosols.

The most important source halogen gases for bromine are the halons and methyl bromide. Halons are used to extinguish fires. Methyl bromide, used as an agricultural fumigant, is another significant source of bromine to the atmosphere. Methane and nitrogen oxide, which react in stratosphere to form water vapour and reactive hydrogen and nitrogen oxides, respectively have ozone depletion capabilities.The emission of oxides of nitrogen by supersonic jet planes is also destructive for ozone layer.

Table: Common Chloroflurocarbons and Halons.

Compound (chemical formula)	Ozone Depletion Potential	Atmospheric Lifetime (Years)	Major Uses
CFC-11 (CFCl3)	1.0	64	Rigid and flexible foams, refrigration
CFC-12 (CF2Cl3)	1.0	108	Air conditioning, refrigeration, rigid foam
CFC-113 (C2F3Cl3)	0.8	88	Solvent
Halon-1211(CF3BrCl)	3.0	25	Portable fire extinguishers
Halon-1301 (CF3Br)	10.0	110	Total flooding fire extinguisher systems
HCFC-22 (HCClF2)	0.05	22	Air conditioning

There are a few halogen source gases present in the stratosphere that have large natural sources. Methyl chloride contributes about 17 per cent of the chlorine currently into the stratosphere. Likewise, about 30 per cent of the bromine is contributed by natural source methyl bromide. These gases are emitted by oceanic and terrestrial ecosystems. Sunspot cycle shows that global total ozone levels vary by 1 to 2 per cent between the maximum and minimum.Volcanoes can emit some chlorine containing gases. In addition, volcanic sulphur and particulate matter emissions reduce solar transmission and ozone abundance. Sunspots and volcanic eruptions bring only short term changes in ozone abundance. Therefore, the main causes of ozone depletion are anthropogenic.

There was no concern about the negative consequences of nitrous oxide and CFCs until three scientists, Paul Crutzen, F.S. Rowland, and Mario Molina, studied the relationship. In 1970, Crutzen highlighted that emissions of nitrous oxide (N_2O), from increased use of fertilizers and supersonic aircrafts could deplete ozone in the stratosphere, where due to photochemical process it is converted into nitric oxide (NO). In 1974, Rowland and Molina concluded that, like N_2O, the CFCs under ultraviolet radiation in stratosphere would be dissociated and released chlorine atoms will destroy ozone layer. This way they alerted the world through their research work that CFCs were probably reducing the average concentration of ozone in the stratosphere.

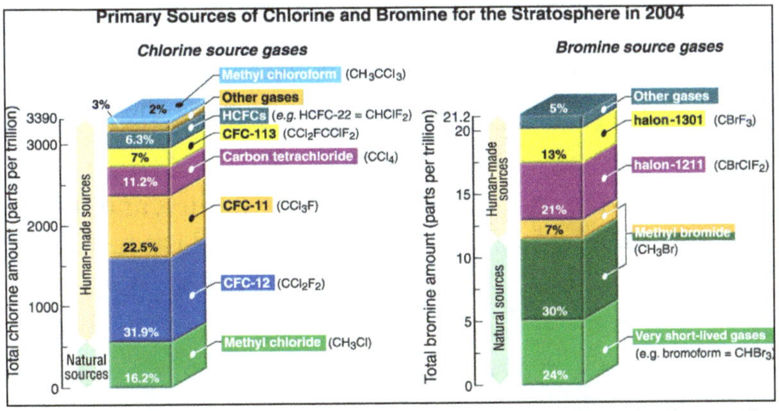

Natural and Anthropogenic Sources of Chlorine and Bromine in Stratosphere.

Later on, when in 1995 ozone hole formation was reported, Dr Rowland expressed his frustration and stated, "What's the use of having developed a science well enough to make predictions, if in the end all we are willing to do is stand around and wait for them come true. Unfortunately, this means that if a disaster is in the making in the stratosphere we are probably not going to avoid it." In 1995, scholarsCrutzen, Rowland and Molina, were awarded the Nobel Prize in chemistry for their pioneering research work on ozone depleting substances.

They discovered that molecules of ozone depleting substances like CFCs drift upward and reach the ozone layer in stratosphere. In stratosphere, CFCs in photochemical process absorb ultraviolet radiation and get decomposed, releasing chlorine. These released chlorine atoms in a complex series of reactions attack molecules of ozone. In a series of chain reactions large numbers of ozone molecules are converted into ordinary oxygen molecules. The chlorine atoms basically interact with ozone molecules by acquiring one oxygen atom to constitute chlorine monoxide (CIO) and leaving behind an oxygen molecule (O_2). As the molecule of chlorine monoxide encounters a single oxygen atom, the oxygen "breaks up" the chlorine monoxide, acquires its oxygen atom and chlorine is released back to indulge in further destruction of ozone in stratosphere. These reactions are chain reactions. In this manner, through chain reactions each chlorine atom that reaches the stratosphere is able to destroy thousands of molecules of ozone resulting into ozone depletion. The simplified reactions are following:

(i) $O_3 + Cl \rightarrow ClO + O_2$ (chlorine acting as the catalyst)

$O + ClO \rightarrow O_2 + Cl$

(ii) $O_3 + NO \rightarrow NO_2 + O_2$ (with nitrous oxide acting as the catalyst)

$O + NO_2 \rightarrow O_2 + NO$

The main cause of ozone depletion and the ozone hole is man-made chemicals, especially man-made halocarbon refrigerants, solvents, propellants, and foam-blowing agents (chlorofluorocarbon (CFCs), HCFCs, halons), referred to as ozone-depleting substances (ODS). These compounds are transported into the stratosphere by the winds after being emitted at the surface. Once in the stratosphere, they release halogen atoms through photodissociation, which catalyze the breakdown of ozone (O_3) into oxygen (O_2). Both types of ozone depletion were observed to increase as emissions of halocarbons increased.

Ozone depletion and the ozone hole have generated worldwide concern over increased cancer risks and other negative effects. The ozone layer prevents most harmful UVB wavelengths of ultraviolet light (UV light) from passing through the Earth's atmosphere. These wavelengths cause skin cancer, sunburn, and cataracts, which were projected to increase dramatically as a result of thinning ozone, as well as harming plants and animals. These concerns led to the adoption of the Montreal Protocol in 1987, which bans the production of CFCs, halons, and other ozone-depleting chemicals.

The ban came into effect in 1989. Ozone levels stabilized by the mid-1990s and began to recover in the 2000s. Recovery is projected to continue over the next century, and the ozone hole is expected to reach pre-1980 levels by around 2075. The Montreal Protocol is considered the most successful international environmental agreement to date.

Ozone Cycle Overview

Three forms (or allotropes) of oxygen are involved in the ozone-oxygen cycle: oxygen atoms (O or atomic oxygen), oxygen gas (O_2 or diatomic oxygen), and ozone gas (O_3 or triatomic oxygen). Ozone is formed in the stratosphere when oxygen molecules photodissociate after intaking ultraviolet photons. This converts a single O_2 into two atomic oxygen radicals. The atomic oxygen radicals then combine with separate O_2 molecules to create two O_3 molecules. These ozone molecules absorb ultraviolet (UV) light, following which ozone splits into a molecule of O_2 and an oxygen atom. The oxygen atom then joins up with an oxygen molecule to regenerate ozone. This is a continuing process that terminates when an oxygen atom recombines with an ozone molecule to make two O_2 molecules.

$$O + O_3 \rightarrow 2 \quad O_2$$

The total amount of ozone in the stratosphere is determined by a balance between photochemical production and recombination.

Ozone can be destroyed by a number of free radical catalysts; the most important are the hydroxyl radical (OH·), nitric oxide radical (NO·), chlorine radical (Cl·) and bromine radical (Br·). The dot is a notation to indicate that each species has an unpaired electron and is thus extremely reactive. All of these have both natural and man-made sources; at the present time, most of the OH· and NO· in the stratosphere is naturally occurring, but human activity has drastically increased the levels of chlorine and bromine. These elements are found in stable organic compounds, especially chlorofluorocarbons, which can travel to the stratosphere without being destroyed in the troposphere due to their low reactivity. Once in the stratosphere, the Cl and Br atoms are released from the parent compounds by the action of ultraviolet light, e.g.

$$CFCl_3 + \text{electromagnetic radiation} \rightarrow Cl\cdot + \cdot CFCl_2$$

Ozone is a highly reactive molecule that easily reduces to the more stable oxygen form with the assistance of a catalyst. Cl and Br atoms destroy ozone molecules through a variety of catalytic cycles. In the simplest example of such a cycle, a chlorine atom reacts with an ozone molecule (O_3), taking an oxygen atom to form chlorine monoxide (ClO) and leaving an oxygen molecule (O_2). The ClO can react with a second molecule of ozone, releasing the chlorine atom and yielding two molecules of oxygen. The chemical shorthand for these gas-phase reactions is:

- $Cl\cdot + O_3 \rightarrow ClO + O_2$

  A chlorine atom removes an oxygen atom from an ozone molecule to make a ClO molecule.

- $ClO + O_3 \rightarrow Cl\cdot + 2\,O_2$

This ClO can also remove an oxygen atom from another ozone molecule; the chlorine is free to repeat this two-step cycle.

The overall effect is a decrease in the amount of ozone, though the rate of these processes can be decreased by the effects of null cycles. More complicated mechanisms have also been discovered that lead to ozone destruction in the lower stratosphere.

A single chlorine atom would continuously destroy ozone (thus a catalyst) for up to two years (the time scale for transport back down to the troposphere) were it not for reactions that remove them from this cycle by forming reservoir species such as hydrogen chloride (HCl) and chlorine nitrate ($ClONO_2$). Bromine is even more efficient than chlorine at destroying ozone on a per atom basis, but there is much less bromine in the atmosphere at present. Both chlorine and bromine contribute significantly to overall ozone depletion. Laboratory studies have also shown that fluorine and iodine atoms participate in analogous catalytic cycles. However, fluorine atoms react rapidly with water and methane to form strongly bound HF in the Earth's stratosphere, while organic molecules containing iodine react so rapidly in the lower atmosphere that they do not reach the stratosphere in significant quantities.

A single chlorine atom is able to react with an average of 100,000 ozone molecules before it is removed from the catalytic cycle. This fact plus the amount of chlorine released into the atmosphere yearly by chlorofluorocarbons (CFCs) and hydrochlorofluorocarbons (HCFCs) demonstrates the danger of CFCs and HCFCs to the environment.

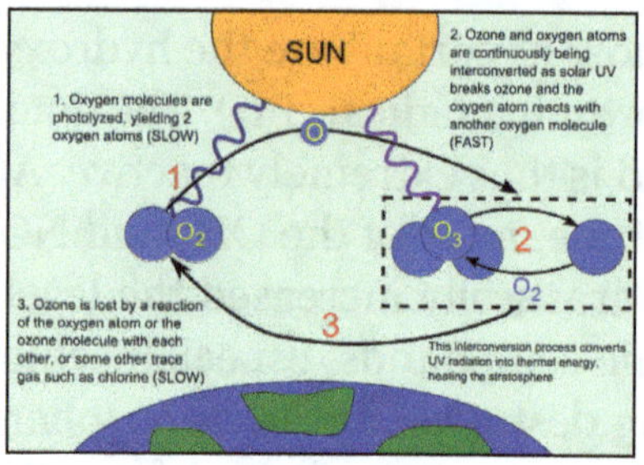

The ozone cycle.

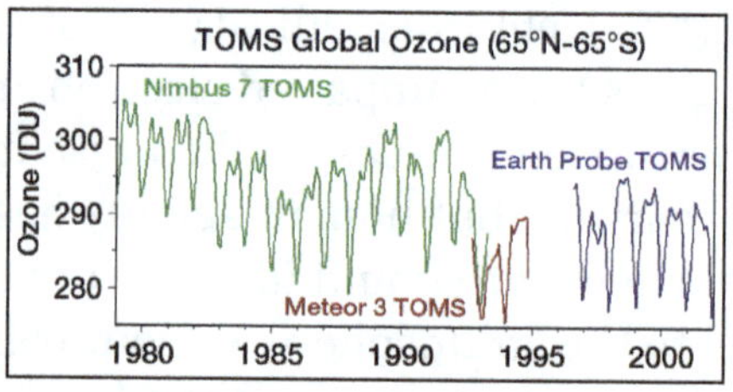

Global monthly average total ozone amount.

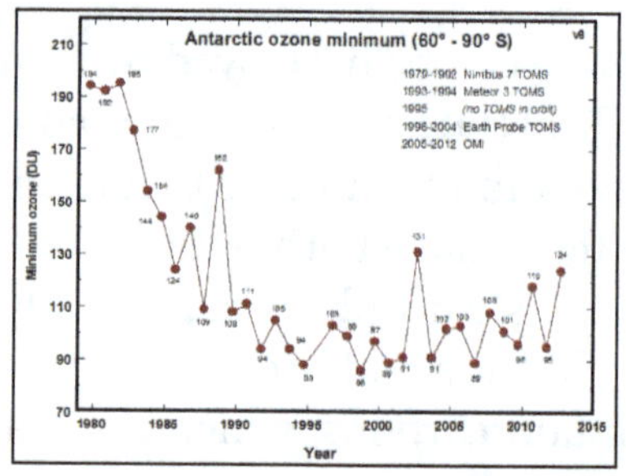

Lowest value of ozone measured by TOMS each year in the ozone hole.

Observations on Ozone Layer Depletion

The ozone hole is usually measured by reduction in the total *column ozone* above a point on the Earth's surface. This is normally expressed in Dobson units; abbreviated as "DU". The most prominent decrease in ozone has been in the lower stratosphere. Marked decreases in column ozone in the Antarctic spring and early summer compared to the early 1970s and before have been observed using instruments such as the Total Ozone Mapping Spectrometer (TOMS).

Reductions of up to 70 percent in the ozone column observed in the austral (southern hemispheric) spring over Antarctica and first reported in 1985 (Farman et al.) are continuing. Antarctic total column ozone in September and October have continued to be 40–50 percent lower than pre-ozone-hole values since the 1990s. A gradual trend toward "healing" was reported in 2016. In 2017, NASA announced that the ozone hole was the weakest since 1988 because of warm stratospheric conditions. It is expected to recover around 2070.

The amount lost is more variable year-to-year in the Arctic than in the Antarctic. The greatest Arctic declines are in the winter and spring, reaching up to 30 percent when the stratosphere is coldest.

Reactions that take place on polar stratospheric clouds (PSCs) play an important role in enhancing ozone depletion. PSCs form more readily in the extreme cold of the Arctic and Antarctic stratosphere. This is why ozone holes first formed, and are deeper, over Antarctica. Early models failed to take PSCs into account and predicted a gradual global depletion, which is why the sudden Antarctic ozone hole was such a surprise to many scientists.

It is more accurate to speak of ozone depletion in middle latitudes rather than holes. Total column ozone declined below pre-1980 values between 1980 and 1996 for mid-latitudes. In the northern mid-latitudes, it then increased from the minimum value by about two percent from 1996 to 2009 as regulations took effect and the amount of chlorine in the stratosphere decreased. In the Southern Hemisphere's mid-latitudes, total ozone remained constant over that time period. There are no significant trends in the tropics, largely because halogen-containing compounds have not had time to break down and release chlorine and bromine atoms at tropical latitudes.

Large volcanic eruptions have been shown to have substantial albeit uneven ozone-depleting effects, as observed with the 1991 eruption of Mt. Pinotubo in the Philippines.

Ozone depletion also explains much of the observed reduction in stratospheric and upper tropospheric temperatures. The source of the warmth of the stratosphere is the absorption of UV radiation by ozone, hence reduced ozone leads to cooling. Some stratospheric cooling is also predicted from increases in greenhouse gases such as CO_2 and CFCs themselves; however, the ozone-induced cooling appears to be dominant.

Predictions of ozone levels remain difficult, but the precision of models' predictions of observed values and the agreement among different modeling techniques have increased steadily. The World Meteorological Organization Global Ozone Research and Monitoring Project—Report No. 44 comes out strongly in favor of the Montreal Protocol, but notes that a UNEP 1994 Assessment overestimated ozone loss for the 1994–1997 period.

Chemicals in the Atmosphere

CFCs and Related Compounds in the Atmosphere

Chlorofluorocarbons (CFCs) and other halogenated ozone depleting substances (ODS) are mainly responsible for man-made chemical ozone depletion. The total amount of effective halogens (chlorine and bromine) in the stratosphere can be calculated and are known as the equivalent effective stratospheric chlorine (EESC).

CFCs were invented by Thomas Midgley, Jr. in the 1930s. They were used in air conditioning and cooling units, as aerosol spray propellants prior to the 1970s, and in the cleaning processes of delicate electronic equipment. They also occur as by-products of some chemical processes. No significant natural sources have ever been identified for these compounds—their presence in the atmosphere is due almost entirely to human manufacture. As mentioned above, when such ozone-depleting chemicals reach the stratosphere, they are dissociated by ultraviolet light to release chlorine atoms. The chlorine atoms act as a catalyst, and each can break down tens of thousands of ozone molecules before being removed from the stratosphere. Given the longevity of CFC molecules, recovery times are measured in decades. It is calculated that a CFC molecule takes an average of about five to seven years to go from the ground level up to the upper atmosphere, and it can stay there for about a century, destroying up to one hundred thousand ozone molecules during that time.

1,1,1-Trichloro-2,2,2-trifluoroethane, also known as CFC-113a, is one of four man-made chemicals newly discovered in the atmosphere by a team at the University of East Anglia. CFC-113a is the only known CFC whose abundance in the atmosphere is still growing. Its source remains a mystery, but illegal manufacturing is suspected by some. CFC-113a seems to have been accumulating unabated since 1960. Between 2010 and 2012, emissions of the gas jumped by 45 percent.

Computer Modeling

Scientists have attributed ozone depletion to the increase of man-made (anthropogenic) halogen compounds from CFCs by combining observational data with computer models. These complex chemistry transport models (e.g. SLIMCAT, CLaMS—Chemical Lagrangian Model of the Stratosphere) work by combining measurements of chemicals and meteorological fields with chemical reaction rate constants. They identify key chemical reactions and transport processes that bring CFC photolysis products into contact with ozone.

Ozone Hole and its Causes

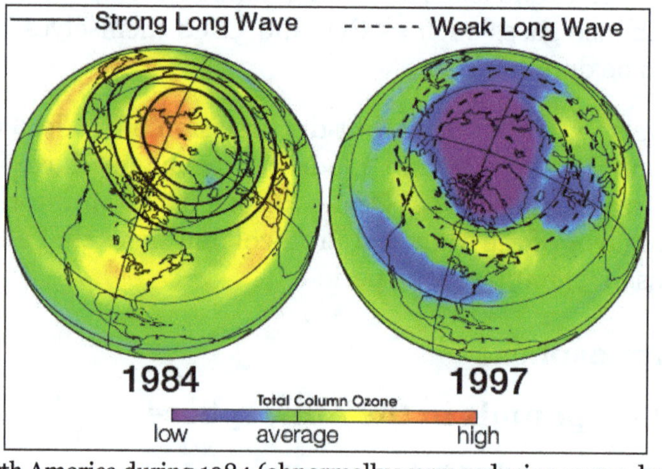

Ozone hole in North America during 1984 (abnormally warm reducing ozone depletion) and 1997 (abnormally cold resulting in increased seasonal depletion).

The Antarctic ozone hole is an area of the Antarctic stratosphere in which the recent ozone levels have dropped to as low as 33 percent of their pre-1975 values. The ozone hole occurs during the Antarctic spring, from September to early December, as strong westerly winds start to circulate around the continent and create an atmospheric container. Within this polar vortex, over 50 percent of the lower stratospheric ozone is destroyed during the Antarctic spring.

As explained above, the primary cause of ozone depletion is the presence of chlorine-containing source gases (primarily CFCs and related halocarbons). In the presence of UV light, these gases dissociate, releasing chlorine atoms, which then go on to catalyze ozone destruction. The Cl-catalyzed ozone depletion can take place in the gas phase, but it is dramatically enhanced in the presence of polar stratospheric clouds (PSCs).

These polar stratospheric clouds (PSC) form during winter, in the extreme cold. Polar winters are dark, consisting of 3 months without solar radiation (sunlight). The lack of sunlight contributes to a decrease in temperature and the polar vortex traps and chills air. Temperatures hover around or below −80 °C. These low temperatures form cloud particles. There are three types of PSC clouds—nitric acid trihydrate clouds, slowly cooling water-ice clouds, and rapid cooling water-ice (nacerous) clouds—provide surfaces for chemical reactions whose products will, in the spring lead to ozone destruction.

The photochemical processes involved are complex but well understood. The key observation is that, ordinarily, most of the chlorine in the stratosphere resides in "reservoir" compounds, primarily chlorine nitrate ($ClONO_2$) as well as stable end products such as HCl. The formation of end products essentially remove Cl from the ozone depletion process. The former sequester Cl, which can be later made available via absorption of light at shorter wavelengths than 400 nm. During the Antarctic winter and spring, however, reactions on the surface of the polar stratospheric cloud particles convert these "reservoir" compounds into reactive free radicals (Cl and ClO). The process by which the clouds remove NO_2 from the stratosphere by converting it to nitric acid in the PSC particles, which then are lost by sedimentation is called denitrification. This prevents newly formed ClO from being converted back into $ClONO_2$.

The role of sunlight in ozone depletion is the reason why the Antarctic ozone depletion is greatest during spring. During winter, even though PSCs are at their most abundant, there is no light over the pole to drive chemical reactions. During the spring, however, the sun comes out, providing energy to drive photochemical reactions and melt the polar stratospheric clouds, releasing considerable ClO, which drives the hole mechanism. Further warming temperatures near the end of spring break up the vortex around mid-December. As warm, ozone and NO_2-rich air flows in from lower latitudes, the PSCs are destroyed, the enhanced ozone depletion process shuts down, and the ozone hole closes.

Most of the ozone that is destroyed is in the lower stratosphere, in contrast to the much smaller ozone depletion through homogeneous gas phase reactions, which occurs primarily in the upper stratosphere.

Interest in Ozone Layer Depletion

Public misconceptions and misunderstandings of complex issues like the ozone depletion are common. The limited scientific knowledge of the public led to a confusion with global warming or the perception of global warming as a subset of the "ozone hole". In the beginning, classical green NGOs refrained from using CFC depletion for campaigning, as they assumed the topic was too complicated. They became active much later, e.g. in Greenpeace's support for a CFC-free fridge produced by the former East German company VEB dkk Scharfenstein.

The metaphors used in the CFC discussion (ozone shield, ozone hole) are not "exact" in the scientific sense. The "ozone hole" is more of a *depression*, less "a hole in the windshield". The ozone does not disappear through the layer, nor is there a uniform "thinning" of the ozone layer. However they resonated better with non-scientists and their concerns. The ozone hole was seen as a "hot issue" and imminent risk as lay people feared severe personal consequences such as skin cancer, cataracts, damage to plants, and reduction of plankton populations in the ocean's photic zone. Not only on the policy level, ozone regulation compared to climate change fared much better in public opinion. Americans voluntarily switched away from aerosol sprays before legislation was enforced, while climate change failed to achieve comparable concern and public action. The sudden recognition in 1985 that there was a substantial "hole" was widely reported in the press. The especially rapid ozone depletion in Antarctica had previously been dismissed as a measurement error. Scientific consensus was established after regulation.

While the Antarctic ozone hole has a relatively small effect on global ozone, the hole has generated a great deal of public interest because:

- Many have worried that ozone holes might start appearing over other areas of the globe, though to date the only other large-scale depletion is a smaller ozone "dimple" observed during the Arctic spring around the North Pole. Ozone at middle latitudes has declined, but by a much smaller extent (about 4–5 percent decrease).

- If stratospheric conditions become more severe (cooler temperatures, more clouds, more active chlorine), global ozone may decrease at a greater pace. Standard global warming theory predicts that the stratosphere will cool.
- When the Antarctic ozone hole breaks up each year, the ozone-depleted air drifts out into nearby regions. Decreases in the ozone level of up to 10 percent have been reported in New Zealand in the month following the breakup of the Antarctic ozone hole, with ultraviolet-B radiation intensities increasing by more than 15 percent since the 1970s.

Consequences of Ozone Layer Depletion

Since the ozone layer absorbs UVB ultraviolet light from the sun, ozone layer depletion increases surface UVB levels (all else equal), which could lead to damage, including increase in skin cancer. This was the reason for the Montreal Protocol. Although decreases in stratospheric ozone are well-tied to CFCs and to increases in surface UVB, there is no direct observational evidence linking ozone depletion to higher incidence of skin cancer and eye damage in human beings. This is partly because UVA, which has also been implicated in some forms of skin cancer, is not absorbed by ozone, and because it is nearly impossible to control statistics for lifestyle changes over time.

Increased UV

Ozone, while a minority constituent in Earth's atmosphere, is responsible for most of the absorption of UVB radiation. The amount of UVB radiation that penetrates through the ozone layer decreases exponentially with the slant-path thickness and density of the layer. When stratospheric ozone levels decrease, higher levels of UVB reach the Earth's surface. UV-driven phenolic formation in tree rings has dated the start of ozone depletion in northern latitudes to the late 1700s.

In October 2008, the Ecuadorian Space Agency published a report called HIPERION. The study used ground instruments in Ecuador and the last 28 years' data from 12 satellites of several countries, and found that the UV radiation reaching equatorial latitudes was far greater than expected, with the UV Index climbing as high as 24 in Quito; the WHO considers 11 as an extreme index and a great risk to health. The report concluded that depleted ozone levels around the mid-latitudes of the planet are already endangering large populations in these areas. Later, the CONIDA, the Peruvian Space Agency, published its own study, which yielded almost the same findings as the Ecuadorian study.

Biological Effects

The main public concern regarding the ozone hole has been the effects of increased surface UV radiation on human health. So far, ozone depletion in most locations has been typically a few percent and, as noted above, no direct evidence of health damage is available in most latitudes. If the high levels of depletion seen in the ozone hole were to be common across the globe, the effects could be substantially more dramatic. As the ozone hole over Antarctica has in some instances grown so large as to affect parts of Australia, New Zealand, Chile, Argentina, and South Africa, environmentalists have been concerned that the increase in surface UV could be significant.

Ozone depletion would magnify all of the effects of UV on human health, both positive (including production of Vitamin D) and negative (including sunburn, skin cancer, and cataracts). In addition, increased surface UV leads to increased tropospheric ozone, which is a health risk to humans.

Basal and Squamous Cell Carcinomas

The most common forms of skin cancer in humans, basal and squamous cell carcinomas, have been strongly linked to UVB exposure. The mechanism by which UVB induces these cancers is well understood—absorption of UVB radiation causes the pyrimidine bases in the DNA molecule to form dimers, resulting in transcription errors when the DNA replicates. These cancers are relatively mild and rarely fatal, although the treatment of squamous cell carcinoma sometimes requires extensive reconstructive surgery. By combining epidemiological data with results of animal studies, scientists have estimated that every one percent decrease in long-term stratospheric ozone would increase the incidence of these cancers by two percent.

Malignant Melanoma

Another form of skin cancer, malignant melanoma, is much less common but far more dangerous, being lethal in about 15–20 percent of the cases diagnosed. The relationship between malignant melanoma and ultraviolet exposure is not yet fully understood, but it appears that both UVB and UVA are involved. Because of this uncertainty, it is difficult to estimate the effect of ozone depletion on melanoma incidence. One study showed that a 10 percent increase in UVB radiation was associated with a 19 percent increase in melanomas for men and 16 percent for women. A study of people in Punta Arenas, at the southern tip of Chile, showed a 56 percent increase in melanoma and a 46 percent increase in nonmelanoma skin cancer over a period of seven years, along with decreased ozone and increased UVB levels.

Cortical Cataracts

Epidemiological studies suggest an association between ocular cortical cataracts and UVB exposure, using crude approximations of exposure and various cataract assessment techniques. A detailed assessment of ocular exposure to UVB was carried out in a study on Chesapeake Bay Watermen, where increases in average annual ocular exposure were associated with increasing risk of cortical opacity. In this highly exposed group of predominantly white males, the evidence linking cortical opacities to sunlight exposure was the strongest to date. Based on these results, ozone depletion is predicted to cause hundreds of thousands of additional cataracts by 2050.

Increased Tropospheric Ozone

Increased surface UV leads to increased tropospheric ozone. Ground-level ozone is generally recognized to be a health risk, as ozone is toxic due to its strong oxidant properties. The risks are particularly high for young children, the elderly, and those with asthma or other respiratory difficulties. At this time, ozone at ground level is produced mainly by the action of UV radiation on combustion gases from vehicle exhausts.

Increased Production of Vitamin D

Vitamin D is produced in the skin by ultraviolet light. Thus, higher UVB exposure raises human vitamin D in those deficient in it. Recent research (primarily since the Montreal Protocol) shows that many humans have less than optimal vitamin D levels. In particular, in the U.S. population, the lowest quarter of vitamin D (<17.8 ng/ml) were found using information from the National Health and Nutrition Examination Survey to be associated with an increase in all-cause mortality in the general population. While blood level of Vitamin D in excess of 100 ng/ml appear to raise blood calcium excessively and to be associated with higher mortality, the body has mechanisms that prevent sunlight from producing Vitamin D in excess of the body's requirements.

Effects on Non-human Animals

A November 2010 report by scientists at the Institute of Zoology in London found that whales off the coast of California have shown a sharp rise in sun damage, and these scientists "fear that the thinning ozone layer is to blame". The study photographed and took skin biopsies from over 150 whales in the Gulf of California and found "widespread evidence of epidermal damage commonly associated with acute and severe sunburn", having cells that form when the DNA is damaged by UV radiation. The findings suggest "rising UV levels as a result of ozone depletion are to blame for the observed skin damage, in the same way that human skin cancer rates have been on the increase in recent decades."

Effects on Crops

An increase of UV radiation would be expected to affect crops. A number of economically important species of plants, such as rice, depend on cyanobacteria residing on their roots for the retention of nitrogen. Cyanobacteria are sensitive to UV radiation and would be affected by its increase. "Despite mechanisms to reduce or repair the effects of increased ultraviolet radiation, plants have a limited ability to adapt to increased levels of UVB, therefore plant growth can be directly affected by UVB radiation."

Public Policy

The full extent of the damage that CFCs have caused to the ozone layer is not known and will not be known for decades; however, marked decreases in column ozone have already been observed. The Montreal and Vienna conventions were installed long before a scientific consensus was established or important uncertainties in the science field were being resolved. The ozone case was understood comparably well by lay persons as e.g. *Ozone shield* or *ozone hole* were useful "easy-to-understand bridging metaphors". Americans voluntarily switched away from aerosol sprays, resulting in a 50 percent sales loss even before legislation was enforced.

After a 1976 report by the United States National Academy of Sciences concluded that credible scientific evidence supported the ozone depletion hypothesis a few countries, including the United States, Canada, Sweden, Denmark, and Norway, moved to eliminate the use of CFCs in aerosol spray cans. At the time this was widely regarded as a first step towards a more comprehensive regulation policy, but progress in this direction slowed in subsequent years, due to a combination of political factors (continued resistance from the halocarbon industry and a general change in attitude towards environmental regulation during the first two years of the Reagan administration) and scientific developments (subsequent National Academy assessments that indicated that the first estimates of the magnitude of ozone depletion had been overly large). A critical DuPont manufacturing patent for Freon was set to expire in 1979. The United States banned the use of CFCs in aerosol cans in 1978. The European Community rejected proposals to ban CFCs in aerosol sprays, and in the U.S., CFCs continued to be used as refrigerants and for cleaning circuit boards. Worldwide CFC production fell sharply after the U.S. aerosol ban, but by 1986 had returned nearly to its 1976 level. In 1993, DuPont Canada closed its CFC facility.

The U.S. Government's attitude began to change again in 1983, when William Ruckelshaus replaced Anne M. Burford as Administrator of the United States Environmental Protection Agency. Under Ruckelshaus and his successor, Lee Thomas, the EPA pushed for an international approach to halocarbon regulations. In 1985 20 nations, including most of the major CFC producers, signed the Vienna Convention for the Protection of the Ozone Layer, which established a framework for negotiating international regulations on ozone-depleting substances. That same year, the discovery of the Antarctic ozone hole was announced, causing a revival in public attention to the issue. In 1987, representatives from 43 nations signed the Montreal Protocol. Meanwhile, the halocarbon industry shifted its position and started supporting a protocol to limit CFC production. However, this shift was uneven with DuPont acting more quickly than their European counterparts. DuPont may have feared court action related to increased skin cancer especially as the EPA had published a study in 1986 claiming that an additional 40 million cases and 800,000 cancer deaths were to be expected in the U.S. in the next 88 years. The EU shifted its position as well after Germany gave up its defence of the CFC industry and started supporting moves towards regulation. Government and industry in France and the UK tried to defend their CFC producing industries even after the Montreal Protocol had been signed.

At Montreal, the participants agreed to freeze production of CFCs at 1986 levels and to reduce production by 50 percent by 1999. After a series of scientific expeditions to the Antarctic produced convincing evidence that the ozone hole was indeed caused by chlorine and bromine from manmade organohalogens, the Montreal Protocol was strengthened at a 1990 meeting in London. The participants agreed to phase out CFCs and halons entirely (aside from a very small amount marked for certain "essential" uses, such as asthma inhalers) by 2000 in non-Article 5 countries and by 2010 in Article 5 (less developed) signatories. At a 1992 meeting in Copenhagen, the phase-out date was moved up to 1996. At the same meeting, methyl bromide (MeBr), a fumigant used primarily in agricultural production, was added to the list of controlled substances. For all substances controlled under the protocol, phaseout schedules were delayed for less developed ('Article 5(1)') countries, and phaseout in these countries was supported by transfers of expertise, technology, and money from non-Article 5(1) Parties to the Protocol. Additionally, exemptions from the agreed schedules could be applied for under the Essential Use Exemption (EUE) process for substances other than methyl bromide and under the Critical Use Exemption (CUE) process for methyl bromide.

Civil society including especially NGOs, played critical roles at all stages of policy development leading up to the Vienna Conference, the Montreal Protocol, and in assessing compliance afterwards. The major companies claimed that no alternatives to HFC existed. An ozone-safe hydrocarbon refrigerant was developed at a Hamburg technological institute in Germany, and in 1992 came to the attention of the non-governmental organization (NGO) Greenpeace. Greenpeace was given the patent, called it "Greenfreeze," and left the patent as open source. The NGO then worked successfully first with a small and struggling company to market an appliance beginning in Europe, then Asia and later Latin America, receiving a 1997 UNEP award. By 1995, Germany had already made CFC refrigerators illegal. Since 2004, corporations like Coca-Cola, Carlsberg, and IKEA have been forming a coalition to promote the ozone-safe Greenfreeze units. Production spread to companies like Electrolux, Bosch, and LG, with sales reaching some 300 million refrigerators by 2008. In Latin America, a domestic Argentinian company began Greenfreeze production in 2003, while the giant Bosch in Brazil began a year later. By 2013 it was being used by some 700 million refrigerators, making up about 40 percent of the market. In the U.S., however, change has been much slower. To some extent, CFCs were being replaced by the less damaging hydrochlorofluorocarbons (HCFCs), although concerns remain regarding HCFCs also. In some applications, hydrofluorocarbons (HFCs) were being used to replace CFCs. HFCs, which contain no chlorine or bromine, do not contribute at all to ozone depletion although they are potent greenhouse gases. The best known of these compounds is probably HFC-134a (R-134a), which in the United States has largely replaced CFC-12 (R-12) in automobile air conditioners. In laboratory analytics (a former "essential" use) the ozone depleting substances can be replaced with various other solvents. Chemical companies like Du Pont, whose representatives even disparaged Greenfreeze as "that German technology," maneuvered the EPA to block the technology in the U.S. until 2011. Ben & Jerry's of Unilever and General Electric, spurred by Greenpeace, had expressed formal interest in 2008 which figured in the EPA's final approval.

More recently, policy experts have advocated for efforts to link ozone protection efforts to climate protection efforts. Many ODS are also greenhouse gases, some thousands of times more powerful agents of radiative forcing than carbon dioxide over the short and medium term. Thus policies protecting the ozone layer have had benefits in mitigating climate change. In fact, the reduction of the radiative forcing due to ODS probably masked the true level of climate change effects of other GHGs, and was responsible for the "slow down" of global warming from the mid-90s. Policy decisions in one arena affect the costs and effectiveness of environmental improvements in the other.

ODS Requirements in the Marine Industry

The IMO has amended MARPOL Annex VI Regulation 12 regarding ozone depleting substances. As from July 1, 2010, all vessels where MARPOL Annex VI is applicable should have a list of equipment using ozone depleting substances. The list should include name of ODS, type and location of equipment, quantity in kg and date. All changes since that date should be recorded in an ODS Record book on board recording all intended or unintended releases to the atmosphere. Furthermore, new ODS supply or landing to shore facilities should be recorded as well.

Prospects of Ozone Depletion

Since the adoption and strengthening of the Montreal Protocol has led to reductions in the emissions of CFCs, atmospheric concentrations of the most-significant compounds have been declining. These substances are being gradually removed from the atmosphere; since peaking in 1994, the Effective Equivalent Chlorine (EECl) level in the atmosphere had dropped about 10 percent by 2008. The decrease in ozone-depleting chemicals has also been significantly affected by a decrease in bromine-containing chemicals. The data suggest that substantial natural sources exist for atmospheric methyl bromide (CH_3Br). The phase-out of CFCs means that nitrous oxide (N_2O), which is not covered by the Montreal Protocol, has become the most highly emitted ozone-depleting substance and is expected to remain so throughout the 21st century.

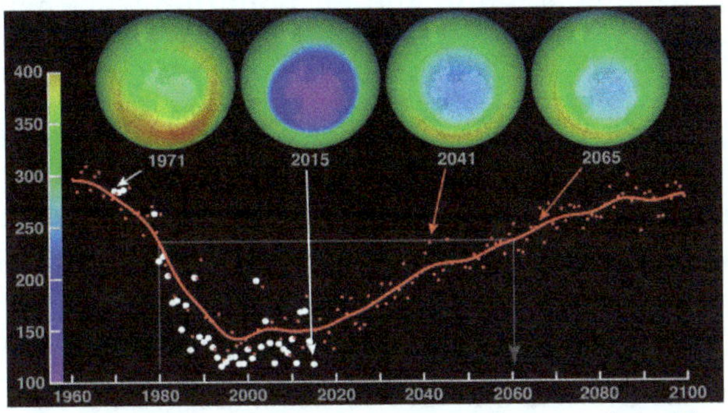

Ozone levels stabilized in the 1990s following the Montreal Protocol, and have started to recover. They are projected to reach pre-1980 levels before 2075.

A 2005 IPCC review of ozone observations and model calculations concluded that the global amount of ozone has now approximately stabilized. Although considerable variability is expected from year to year, including in polar regions where depletion is largest, the ozone layer is expected to begin to recover in coming decades due to declining ozone-depleting substance concentrations, assuming full compliance with the Montreal Protocol.

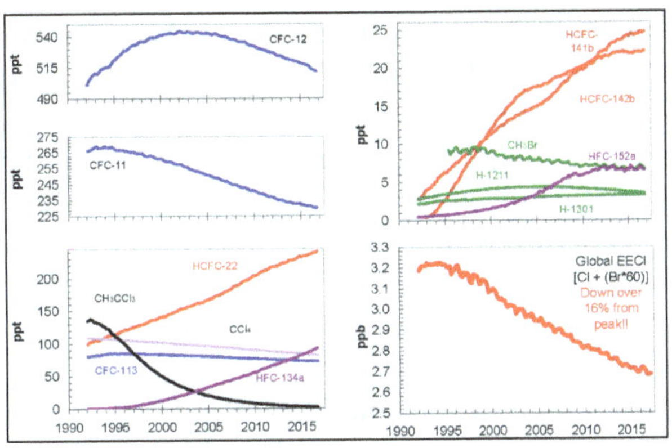

Ozone-depleting gas trends.

The Antarctic ozone hole is expected to continue for decades. Ozone concentrations in the lower stratosphere over Antarctica will increase by 5–10 percent by 2020 and return to pre-1980 levels by about 2060–2075. This is 10–25 years later than predicted in earlier assessments, because of revised estimates of atmospheric concentrations of ozone-depleting substances, including a larger predicted future usage in developing countries. Another factor that may prolong ozone depletion is the drawdown of nitrogen oxides from above the stratosphere due to changing wind patterns. A gradual trend toward "healing" was reported in 2016.

Ozone Depletion and Climate Change

Ozone depletion and climate change, or Ozone hole and global warming in more popular terms, are environmental challenges whose connections have been explored and which have been compared and contrasted, for example in terms of global regulation, in various studies and books.

There is widespread scientific interest in better regulation of climate change, ozone depletion and air pollution, as in general the human relationship with the biosphere is deemed of major historiographical and political significance. Already by 1994 the legal debates about respective regulation regimes on climate change, ozone depletion and air pollution were being dubbed "monumental" and a combined synopsis provided.

There are some parallels between atmospheric chemistry and anthropogenic emissions in the discussions which have taken place and the regulatory attempts which have been made. Most important is that the gases causing both problems have long lifetimes after emission to the atmosphere, thus causing problems which are difficult to reverse. However, the Vienna Convention for the Protection of the Ozone Layer and the Montreal Protocol that amended it are seen as success stories, while the Kyoto Protocol on anthropogenic climate change has largely failed. Currently efforts are being undertaken to assess the reasons and to use synergies, for example with regard to data reporting and policy design and further exchanging of information. While the general public tends to see global warming as a subset of ozone depletion, in fact ozone and chemicals such as chlorofluorocarbons (CFCs) and other halocarbons, which are held responsible for ozone depletion, are important greenhouse gasses. Furthermore, natural levels of ozone in both the stratosphere and troposphere have a warming effect.

Policy Approach

There are both links and major differences between ozone depletion and global warming and the way the two challenges have been handled. While in the case of atmospheric ozone depletion, in a situation of high uncertainty and against strong resistance, climate change regulation attempts at the international level such as the Kyoto Protocol have failed to reduce global emissions. The Vienna Convention for the Protection of the Ozone Layer and the Montreal Protocol were both originally signed by only some member states of the United Nations (43 nations in the case of the Montreal Protocol in 1986) while Kyoto attempted to create a worldwide agreement from scratch. Expert consensus concerning CFCs in the form of the Scientific Assessment of Ozone Depletion was reached long after the first regulatory steps were taken, and as of 29 December 2012, all countries in the United Nations plus the Cook Islands, the Holy See, Niue and the supranational European Union had ratified the original Montreal Protocol. These countries have also ratified the London, Copenhagen, and Montreal amendments to the Protocol. As of 15 April 2014, the Beijing amendments had not been ratified by two state parties.

After the Vienna Convention, the halocarbon industry shifted its position and started supporting a protocol to limit CFC production. US manufacturer DuPont acted more quickly than their European counterparts. The EU shifted its position as well after Germany, which has a substantial chemical industry, gave up its defence of the CFC industry and started supporting more regulation. Government and industry in France and the UK had tried to defend their CFC-producing industries even after the Montreal Protocol had been signed.

The Vienna Convention was installed before a scientific consensus on the ozone hole was established. On the contrary, until the 1980s the EU, NASA, NAS, UNEP, WMO and the British government had issued scientific reports with divergent conclusions. Sir Robert (Bob) Watson, Director of the Science Division at NASA, played a crucial role in the process of reaching a unified assessment.

Policy and Consensus

Bob Watson, successfully united the international science community in 1985 to act on the problem with the ozone hole, before a consensus existed.

Sir Robert (Bob) Watson played an important role in both cases.

Aant Elzinga wrote in 1996 about the consensus, that the Intergovernmental Panel on Climate Change has tried in the prior two reports a global consensus approach to climate action. Stephen Schneider and Paul N. Edwards, noted in 1997, that after the IPCC Second Assessment Report, the lobby group Global Climate Coalition and a few self-proclaimed "contrarian" scientists tried to discredit the conclusions of the report. They pointed out that the goal of the IPCC is to fairly represent the complete range of credible scientific opinion and if possible a consensus view.

In 2007, Reiner Grundmann compared climate actions in Europe and the United States, he interpreted the inaction besides existing consensus, and noted, *Political agenda that drove US climate change policy. The high visibility of sceptical scientists in the media resonates with this*, and wrote that Germany started ambitious goals, reduced emissions, because 'balanced reporting' led to a bias in climate change coverage in advantage of sceptical arguments in the U.S., but not so much in Germany. Additionally, Grundmann pointed out that after warnings from scientists in 1986 the German Parliament commissioned the *Enquetekommission 'Vorsorge zum Schutz der Erdatmosphäre'* (Precaution for the Protection of the Earth's Atmosphere), to assess the situation, consisting of scientists, politicians and representatives of interest groups. Three years later the report made an impact with the assessment of the state of the art in climate research, an assessment of the threat of climate change itself as well as suggestions for clear emissions reduction targets, even though he argues there was no consensus, and attributed the success of the report to strong precautionary action, and that no scientific outsiders or climate change skeptics were involved.

A linear model of policy-making, based on a position that "the more knowledge we have, the better the political response will be", was not applied in the ozone case. On the contrary, the CFC regulation process focused more on managing ignorance and uncertainties as a basis of political decision making, as the relationships between science, public (lack of) understanding and policy were better taken into account. In the meantime, such a player in the IPCC process as Michael

Oppenheimer conceded some limitations of the IPCC consensus approach and asked for concurring, smaller assessments of special problems instead of repetitions of the large-scale approach every six years. It has become more important to provide a broader exploration of uncertainties. Others also see mixed blessings in the drive for consensus within the IPCC process and have asked for dissenting or minority positions to be included or for statements about uncertainties to be improved.

Public Opinion

The two atmospheric problems have achieved significantly different levels of understanding by the public, including both the basic science and policy issues. People have limited scientific knowledge about global warming and tend to confuse it with or see it as a subset of the ozone hole. Not only on the policy level, ozone regulation fared much better than climate change in public opinion. Americans voluntarily switched away from aerosol sprays before legislation was enforced, while climate change has failed in achieving a broader scientific comprehension and in raising comparable concern.

The metaphors used in the CFC discussion (ozone shield, ozone hole) resonated better with non-scientists and their concerns. The ozone case was communicated to lay persons "with easy-to-understand bridging metaphors derived from the popular culture" and related to "immediate risks with everyday relevance", while the public opinion on climate change sees no imminent danger. The ozone hole was much more seen as a "hot issue" and imminent risk compared to global climate change, as lay people feared a depletion of the ozone layer (ozone shield) risked increasing severe consequences such skin cancer, cataracts, damage to plants, and reduction of plankton populations in the ocean's photic zone. This was not the case with global warming.

Personal Risk Assessment and Knowledge

Sheldon Ungar, a Canadian sociologist, assumes that while the quantity of specialized knowledge is exploding, in contrast scientific ignorance among lay people is the norm and even increasing. Public opinion failed to tie climate change to concrete events which could be used as a threshold or beacon to signify immediate danger. Scientific predictions of a temperature rise of 2 °C (4 °F) to 3 °C (5 °F) over several decades do not resonate with people, for example in North America, who experience similar swings during a single day. As scientists define global warming as a problem of the future, a liability in the "attention economy", pessimistic outlooks in general and the attribution of extreme weather to climate change have often been discredited or ridiculed in the public arena (compare the Gore effect). Even when James Hansen tried to use the 1988–89 North American drought as a call to action, scientists kept stating, in line with the IPCC findings, that even extreme weather is not climate. While the greenhouse effect, *per se*, is essential for life on earth, the case was quite different with the ozone hole and other metaphors about ozone depletion. The scientific assessment of the ozone problem also had large uncertainties; both the ozone content of the upper atmosphere and its depletion are complicated to measure and the link between ozone depletion and rates of enhanced skin cancer is rather weak. But the metaphors used in the discussion (ozone shield, ozone hole) resonated better with lay people and their concerns.

> The idea of rays penetrating a damaged "shield" meshes nicely with abiding and resonant cultural motifs, including "Hollywood affinities." These range from the shields on the Starship Enterprise to Star Wars... It is these pre-scientific bridging metaphors built

> around the penetration of a deteriorating shield that render the ozone problem relatively simple. That the ozone threat can be linked with Darth Vader means that it is encompassed in common sense understandings that are deeply ingrained and widely shared.
>
> —*Sheldon Ungar*

The CFC regulation attempts at the end of the 1980s profited from those easy to grasp metaphors and the personal risk assumptions taken from them. The fate of celebrities like President Ronald Reagan, who had skin cancer removal from his nose in 1985 and 1987, was also of high importance. In case of the public opinion on climate change, no imminent danger is perceived.

Cost-benefit Assessments and Industry Policy

Cass Sunstein and others have compared the differing approach of the United States to the Montreal Protocol, which it accepted, and the Kyoto Protocol, which it rejected. Sunstein assumes that the cost-benefit assessments of climate change action for the US were instrumental in the US' withdrawal from participation in Kyoto. Daniel Magraw, also a lawyer, considers governmental motivations besides relative costs and benefits as being of higher importance. Peter Orszag and Terry Dinan took an insurance perspective and assume that an assessment which predicted dire consequences of climate change would be more of a motivation for the US to change its stance on global warming and adopting regulation measurements.

The US chemical company DuPont had already lost some of their zeal in defending their products after a strategic manufacturing patent for Freon was set to expire in 1979. A citizen boycott of spray cans gained importance in parallel. Not by chance, the United States banned the use of CFCs in aerosol cans in 1978.

Government and industry in France and the UK tried to defend their CFC-producing industries even after the Montreal Protocol had been signed. The European Community rejected proposals to ban CFCs in aerosol sprays for a long time. The EU shifted its position after Germany, which also has a large chemical industry, gave up its defence of the CFC industry and started supporting moves towards regulation. After regulation was more and more enforced, DuPont acted faster than their European counterparts as they may have feared court action related to increased skin cancer, especially as the EPA had published a study in 1986 claiming that an additional 40 million cases and 800,000 cancer deaths were to be expected in the US in the next 88 years. The identification and marketing of a 100% ozone-safe hydrocarbon refrigerant called "Greenfreeze" by the NGO Greenpeace in the early 1990s had a rapid significant impact in major markets of Europe and Asia. The climate change protocols were less successful. In the case of Kyoto, then secretary of the environment Angela Merkel, prevented a possible failure by suggesting to use 1990 as starting date for emission reduction. In so far the demise of the Eastern European heavy industry allowed for a high commitment, but actual emissions kept on growing on a global scale.

Science Background

There are various links between the two fields of human-atmospheric interaction. Policy experts have advocated for a closer linking of ozone protection and climate protection efforts.

Drew Shindell has used climate models to assess both climate change and ozone depletion. In his view, while research up to now has been more about the impact of CFC emissions on stratospheric ozone, the future will be more about the interaction between climate change and ozone feedback. Ozone is a greenhouse gas itself. Many ozone-depleting substances are also greenhouse gases, some agents of radiative forcing are thousands of times more powerful than carbon dioxide over the short and medium term. The increases in concentrations of these chemicals have produced 0.34 ± 0.03 W/m^2 of radiative forcing, corresponding to about 14% of the total radiative forcing from increases in the concentrations of well-mixed greenhouse gases. Already the natural ozone variability in the stratosphere seems to be closely correlated with the 11-year solar cycle of irradiance changes and has, via a dynamic coupling between the stratosphere and troposphere, a significant impact on climate.

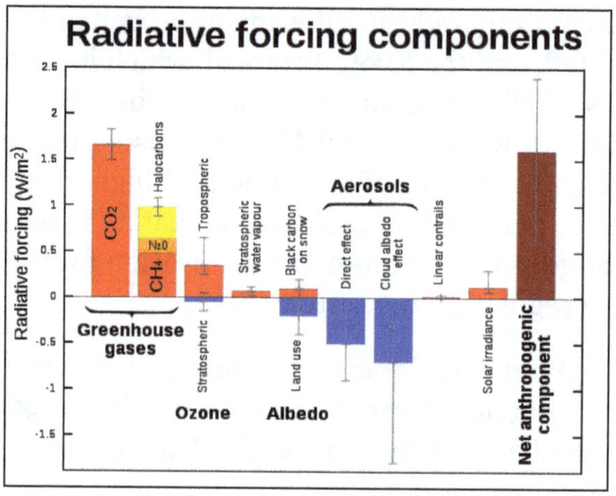

Radiative forcing from various greenhouse gases.

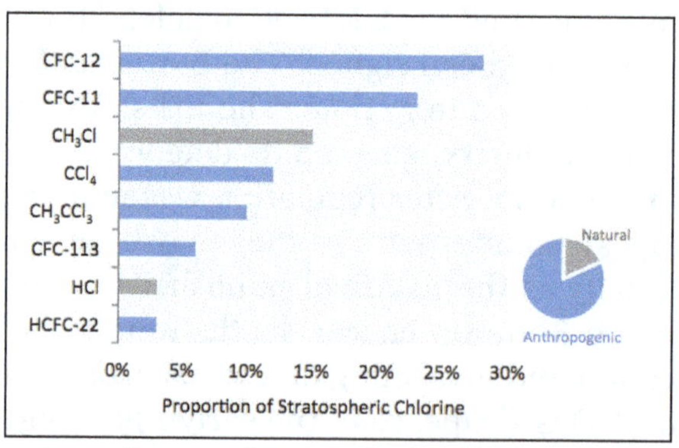

Sources of Stratospheric Chlorine.

As with carbon dioxide and methane, there are some natural sources of tropospheric chlorine, such as sea spray. Chlorine from ocean spray is soluble and thus is washed by rainfall before it reaches the stratosphere. It is *stratospheric* chlorine that affects ozone depletion. Only methyl chloride, which is one of the halocarbons, has a mainly natural source, and it is responsible for about 20% of the chlorine in the stratosphere; the remaining 80% comes from man-made sources. Chlorofluorocarbons, in contrast, are insoluble and long-lived, allowing them to reach the stratosphere. In the lower atmosphere, there is much more chlorine from CFCs and related haloalkanes than there is in hydrogen chloride from salt spray, and in the stratosphere halocarbons are dominant.

The same CO_2 radiative forcing that produces global warming is expected to cool the stratosphere. This cooling, in turn, is expected to produce a relative *increase* in ozone (O_3) depletion in the polar area and in the frequency of ozone holes. Conversely, ozone depletion represents a radiative forcing of the climate system of about -0.15 ± 0.10 watts per square metre (W/m^2).

References

- "Stratospheric Ozone and Surface Ultraviolet Radiation". Scientific Assessment of Ozone Depletion: 2010 (PDF). WMO. 2011. Retrieved 01, February 2020
- Grundmann, Reiner (2007). "Climate Change and Knowledge Politics" (PDF). Environmental Politics. 16 (3): 414–432. doi:10.1080/09644010701251656. Archived from the original(PDF) on August 26, 2014. Retrieved 15, April 2020
- Grundmann, Reiner (2001). Transnational Environmental Policy : the ozone layer. New York: Routledge. ISBN 0-415-22423-3
- Tabin, Shagoon (2008). Global Warming: The Effect Of Ozone Depletion. APH Publishing. p. 194. ISBN 9788131303962. Retrieved 21, June 2020
- Solomon, Susan, et al. (June 30, 2016). "Emergence of healing in the Antarctic ozone layer". Science. 353 (6296): 269–74. Bibcode:2016Sci...353..269S. doi:10.1126/science.aae0061. PMID 27365314
- Ungar, Sheldon (2000). "Knowledge, ignorance and the popular culture: climate change versus the ozone hole". Public Understanding of Science. 9 (3): 297–312. doi:10.1088/0963-6625/9/3/306. Retrieved 18, March 2020

Chapter 5

Climate Change: Evidence, Causes and Impacts

Climate change is the global phenomenon of climate transformation characterized by long term changes in the weather patterns that have defined the Earth's local, regional and global climates. Activities like burning of fossil fuels and deforestation have contributed to this change. Melting of the polar ice caps, rising sea levels and global warming are some of the negative impacts of climate change. This chapter is dedicated to addressing this urgent issue that threatens the very survival of life on Earth.

The Earth's average temperature is about 15 °C but has been much higher and lower in the past. There are natural fluctuations in the climate but scientists say temperatures are now rising faster than at many other times.

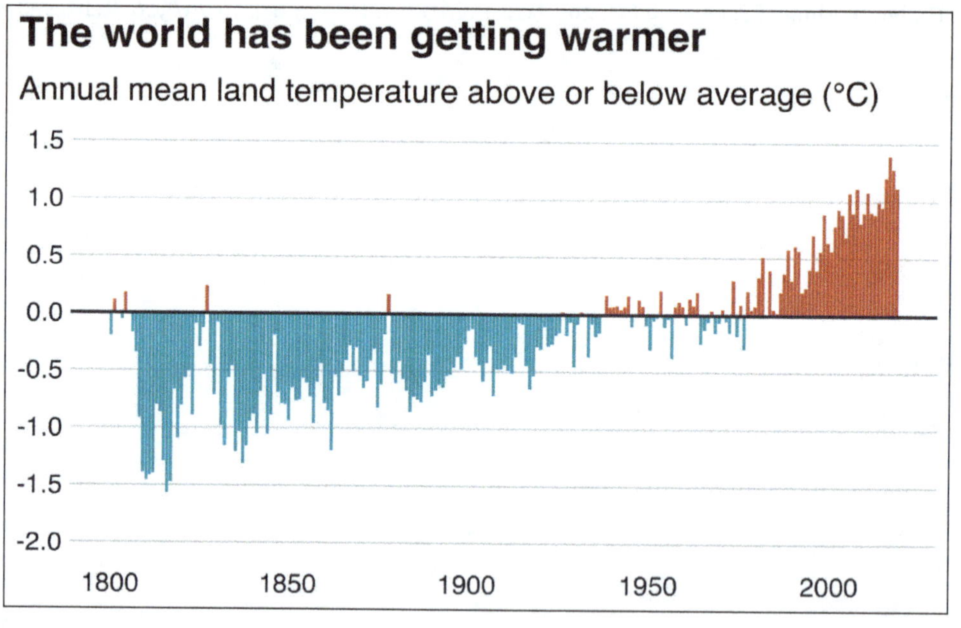

This is linked to the greenhouse effect, which describes how the Earth's atmosphere traps some of the Sun's energy. Solar energy radiating back to space from the Earth's surface is absorbed by greenhouse gases and re-emitted in all directions. This heats both the lower atmosphere and the surface of the planet. Without this effect, the Earth would be about 30C colder and hostile to life.

The Greenhouse Effect

Scientists believe we are adding to the natural greenhouse effect, with gases released from industry and agriculture trapping more energy and increasing the temperature. This is known as climate change or global warming.

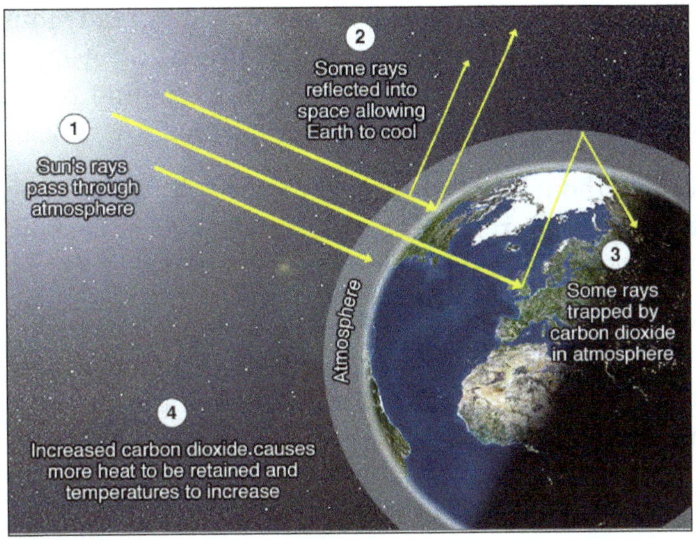

What are Greenhouse Gases?

The greenhouse gas with the greatest impact on warming is water vapor. But it remains in the atmosphere for only a few days. Carbon dioxide (CO_2), however, persists for much longer. It would take hundreds of years for a return to pre-industrial levels and only so much can be soaked up by natural reservoirs such as the oceans. Most man-made emissions of CO_2 come from burning fossil fuels. When carbon-absorbing forests are cut down and left to rot, or burned, that stored carbon is released, contributing to global warming.

Since the Industrial Revolution began in about 1750, CO_2 levels have risen more than 30%. The concentration of CO_2 in the atmosphere is higher than at any time in at least 800,000 years. Other greenhouse gases such as methane and nitrous oxide are also released through human activities but they are less abundant than carbon dioxide.

What is the Evidence for Warming?

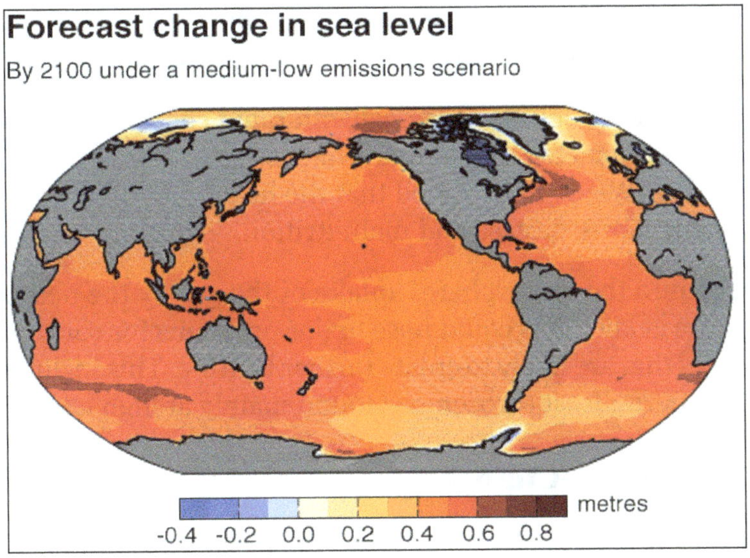

The world is about one degree Celsius warmer than before widespread industrialization, the World Meteorological Organization (WMO). It says the past five years, 2015–2019, were the warmest on record. Across the globe, the average sea level increased by 3.6mm per year between 2005 and 2015. Most of this change was because water increases in volume as it heats up.

However, melting ice is now thought to be the main reason for rising sea levels. Most glaciers in temperate regions of the world are retreating. And satellite records show a dramatic decline in Arctic sea-ice since 1979. The Greenland Ice Sheet has experienced record melting in recent years.

Satellite data also shows the West Antarctic Ice Sheet is losing mass. A recent study indicated East Antarctica may also have started to lose mass. The effects of a changing climate can also be seen in vegetation and land animals. These include earlier flowering and fruiting times for plants and changes in the territories of animals.

How much will Temperatures Rise in Future?

The change in the global surface temperature between 1850 and the end of the 21st Century is likely to exceed 1.5 °C, most simulations suggest. The WMO says that if the current warming trend continues, temperatures could raise 3-5 °C by the end of this century. Temperature rises of 2 °C had long been regarded as the gateway to dangerous warming. More recently, scientists and policymakers have argued that limiting temperature rises to 1.5 °C is safer.

How will Climate Change Affect us?

There is uncertainty about how great the impact of a changing climate will be. It could cause fresh water shortages, dramatically alter our ability to produce food, and increase the number of deaths from floods, storms and heat waves. This is because climate change is expected to increase the frequency of extreme weather events - though linking any single event to global warming is complicated.

As the world warms, more water evaporates, leading to more moisture in the air. This means many areas will experience more intense rainfall - and in some places snowfall. But the risk of drought in inland areas during hot summers will increase. More flooding is expected from storms and rising sea levels. But there are likely to be very strong regional variations in these patterns.

Poorer countries, which are least equipped to deal with rapid change, could suffer the most. Plant and animal extinctions are predicted as habitats change faster than species can adapt. And the World Health Organization (WHO) has warned that the health of millions could be threatened by increases in malaria, water-borne disease and malnutrition.

As more CO_2 is released into the atmosphere, uptake of the gas by the oceans increases, causing the water to become more acidic. This could pose major problems for coral reefs. Global warming will cause further changes that are likely to create further heating. This includes the release of large quantities of methane as permafrost - frozen soil found mainly at high latitudes - melts.

Natural Sources of Climate Change

Some of the factors mentioned below are responsible for long term climate change whereas others are responsible for short term climate change. Major natural sources of climate change are solar variability, changes in Earth's orbit and tilt, plate tectonics and biological evolution.

Solar Variability

This happens due to variation in solar energy output. The energy output of the sun varies slightly over time. This has been measured by the data obtained through satellite. The variation in the total output has been observed both throughout the day as well as during the solar cycle. These variations can directly affect Earth's climate. But the exact role that solar variability plays is still a controversial and not unanimously accepted by the scientific community.

Changes in Earth's Orbit and Tilt

This was another major natural source responsible for climate change. You might have read that earth generally changes its orbit over a 100,000- year cycle. It normally shifts from almost a circular to elliptical orbit. This has led to changes in the tilt of Earth's axis between 21.8 and 24.4 degrees. There has been change in the planet's solar energy budget. In other words, there used to be a change in the amount of solar radiation that different areas of the Earth receive. The northern hemisphere is now closer to the sun in winter and farther away in summer. It receives five percent less sunlight in summer than 12,000 years ago. More tilt leads to greater seasonal variations. This is known as Milankovitch cycle. Milankovitch was a Serbian geophysicist and astronomer who theorized this above mentioned phenomena.

Figure: Climate Change due toChanges in Earth's Orbit and Tilt.

Plate Tectonics

Millions of years ago most of the Earth's land mass was a single continent known as Pangea and surrounded by one huge water bodies known as panthalas. Over time, portions of it drifted apart, gradually forming the continents and ocean basins existing today. The redistribution of land mass and ocean area has had a major effect on global climate.

Biological Evolution

This natural process took place mainly during the evolution of photosynthetic organism. Several billion years ago Earth's pre-life atmosphere was mainly covered by carbon dioxide and methane? The gradual shift in the atmospheric concentrations from carbon dioxide and methane to nitrogen and oxygen took place because photosynthetic organisms consumed carbon dioxide and generated oxygen.

Modeling and Projection of Anthropogenic Climate Change

The behaviour of the climate system, its components and their interactions, can be studied and simulated using tools known as climate models. These are designed mainly for studying climate processes and natural climate variability, and for projecting the response of the climate to human-induced forcing. Each component or coupled combination of components of the climate system can be represented by models of varying complexity.

The nucleus of the most complex atmosphere and ocean models, called General Circulation Models (Atmospheric General Circulation Models (AGCMs) and Ocean General Circulation Models (OGCMs)) is based upon physical laws describing the dynamics of atmosphere and ocean, expressed by mathematical equations. Since these equations are non-linear, they need to be solved numerically by means of well-established mathematical techniques. Current atmosphere models are solved spatially on a three-dimensional grid of points on the globe with a horizontal resolution typically of 250 km and some 10 to 30 levels in the vertical. A typical ocean model has a horizontal resolution of 125 to 250 km and a resolution of 200 to 400 m in the vertical. Their time-dependent behaviour is computed by taking time steps typically of 30 minutes.

Models of the various components of the climate system may be coupled to produce increasingly complex models. Processes taking place on spatial and temporal scales smaller than the model's resolution, such as individual clouds or convection in atmosphere models, or heat transport through boundary currents or mesoscale eddies in ocean models are included through a parametric representation in terms of the resolved basic quantities of the model. Coupled atmosphere-ocean models, including such parametrized physical processes, are called Atmosphere-Ocean General Circulation Models (AOGCMs). They are combined with mathematical representations of other components of the climate system, sometimes based on empirical relations, such as the land surface and the cryosphere. The most recent models may include representations of aerosol processes and the carbon cycle, and in the near future perhaps also the atmospheric chemistry. The development of these very complex coupled models goes hand in hand with the availability of ever larger and faster computers to run the models. Climate simulations require the largest, most capable computers available.

A realistic representation of the coupling between the various components of the climate system is essential. In particular the coupling between the atmosphere and the oceans is of central importance. The oceans have a huge heat capacity and a decisive influence on the hydrological cycle of the climate system, and store and exchange large quantities of carbon dioxide. To a large degree the coupling between oceans and atmosphere determines the energy budget of the climate system. There have been difficulties modelling this coupling with enough accuracy to prevent the model climate unrealistically drifting away from the observed climate. Such climate drift may be avoided by adding an artificial correction to the coupling, the socalled "flux adjustment". The evaluation of recent model results identifies improvements since the SAR, to the point that there is a reduced reliance on such corrections, with some recent models operating with minimal or no adjustment.

For various reasons, it is important to also develop and use simpler models than the fully coupled comprehensive AOGCMs, for example to study only one or a specific combination of components of the climate system or even single processes, or to study many different alternatives, which is not possible or is impractical with comprehensive models. In IPCC (1997) a hierarchy of models used in the IPCC assessment process was identified and described, differing in such aspects as the number of spatial dimensions, the extent to which physical processes are explicitly represented, the level to which empirical parameterization is involved, and the computational costs of running the models. In the IPCC context, simple models are also used to compute the consequences of greenhouse gas emission scenarios. Such models are tuned to the AOGCMs to give similar results when globally averaged.

Projections of Climate Change

Climate models are used to simulate and quantify the climate response to present and future human activities. The first step is to simulate the present climate for extended simulation periods, typically many decades, under present conditions without any change in external climate forcing.

The quality of these simulations is assessed by systematically comparing the simulated climate with observations of the present climate. In this way the model is evaluated and its quality established. A range of diagnostic tools has been developed to assist the scientists in carrying out the evaluation. This step is essential to gain confidence in and provide a baseline for projections of human-induced climate change. Models may also be evaluated by running them under different palaeoclimate (e.g. Ice Age) conditions. Once the quality of the model is established, two different strategies have been applied to make projections of future climate change.

The first, so-called equilibrium method is to change, e.g. double, the carbon dioxide concentration and to run the model again to a new equilibrium. The differences between the climate statistics of the two simulations provide an estimate of the climate change corresponding to the doubling of carbon dioxide, and of the sensitivity of the climate to a change in the radiative forcing. This method reduces systematic errors present in both simulations. If combined with simple slab ocean models, this strategy is relatively cheap because it does not require long runs to reach equilibrium. However it does not provide insight in to the time dependence of climate change.

The second, so-called transient, method, common nowadays with improved computer resources, is to force the model with a greenhouse gas and aerosol scenario. The difference between such simulation and the original baseline simulation provides a time-dependent projection of climate change. This transient method requires a time-dependent profile of greenhouse gas and aerosol concentrations. These may be derived from so-called emission scenarios. Such scenarios have been developed, among others by IPCC, on the basis of various internally coherent assumptions concerning future socioeconomic and demographic developments. In the SAR the IPCC Scenarios IS92 were used. The most recent IPCC emission scenarios are described in the IPCC Special Report on Emission Scenarios. Different assumptions concerning e.g. the growth of the world population, energy intensity and efficiency, and economic growth, lead to considerably different emission scenarios. For example the two extreme estimates in the IPCC IS92 scenarios of the carbon dioxide emission by 2100 differ by a factor of 7. Because scenarios by their very nature should not be used and regarded as predictions, the term "climate projections" is used in this Report.

Transient simulations may also be based on artificially constructed, so-called idealised, scenarios.

For example, scenarios have been constructed, assuming a gradual increase of greenhouse gas concentrations followed by stabilization at various levels. Climate simulations based on such idealised scenarios may provide insight in to the climate response to potential policy measures leading to a stabilisation of the GHG concentrations, which is the ultimate objective of the United Nations Framework Convention on Climate Change (UNFCCC) as formulated in its Article.

Projections from present models show substantial agreement, but at the same time there is still a considerable degree of ambiguity and difference between the various models. All models show an increase in the globally averaged equilibrium surface temperature and global mean precipitation.

Model results are more ambiguous about the spatial patterns of climate change than about the global response. Regional patterns depend significantly on the time dependence of the forcing, the spatial distribution of aerosol concentrations and details of the modelled climate processes. Research tools have been developed to generate more reliable regional climate information.

To study the impact of climate change, a plausible and consistent description of a possible future climate is required. The construction of such climate change scenarios relies mainly on results from model projections, although sometimes information from past climates is used.

Predictability, Global and Regional

In trying to quantify climate change, there is a fundamental question to be answered: is the evolution of the state of the climate system predictable? Since the pioneering work by Lorenz in the 1960s, it is well known that complex non-linear systems have limited predictability, even though the mathematical equations describing the time evolution of the system are perfectly deterministic.

The climate system is, as we have seen, such a non-linear complex system with many inherent time scales. Its predictability may depend on the type of climate event considered, the time and space scales involved and whether internal variability of the system or variability from changes in external forcing is involved. Internal variations caused by the chaotic dynamics of the climate system may be predictable to some extent. Recent experience has shown that the ENSO phenomenon may possess a fair degree of predictability for several months or even a year ahead. The same may be true for other events dominated by the long oceanic timescales, such as perhaps the NAO. On the other hand, it is not known, for example, whether the rapid climate changes observed during the last glacial period are at all predictable or are unpredictable consequences of small changes resulting in major climatic shifts.

There is evidence to suggest that climate variations on a global scale resulting from variations in external forcing are partly predictable. Examples are the mean annual cycle and short-term climate variations from individual volcanic eruptions, which models simulate well. Regularities in past climates, in particular the cyclic succession of warm and glacial periods forced by geometrical changes in the Sun-Earth orbit, are simulated by simple models with a certain degree of success. The global and continental scale aspects of human-induced climate change, as simulated by the models forced by increasing greenhouse gas concentration, are largely reproducible. Although this is not an absolute proof, it provides evidence that such externally forced climate change may be predictable, if their forcing mechanisms are known or can be predicted. Finally, global or continental scale climate change and variability may be more predictable than regional or local scale change, because the climate on very large spatial scales is less influenced by internal dynamics, whereas regional and local climate is much more variable under the influence of the internal chaotic dynamics of the system.

Rapid Climate Change

A non-linear system such as the climate system may exhibit rapid climate change as a response to internal processes or rapidly changing external forcing. Because the probability of their occurrence may be small and their predictability limited, they are colloquially referred to as "unexpected events" or "surprises". The abrupt events that took place during the last glacial cycle are often cited as an example to demonstrate the possibility of such rapid climate change. Certain possible abrupt events as a result of the rapidly increasing anthropogenic forcing could be envisioned. Examples are a possible reorganization of the thermohaline ocean circulation in the North Atlantic resulting in a more southerly course of the Gulf Stream, which would have a profound influence on the climate of Western Europe, a possible reduction of upper-level ocean cycling in the Southern Ocean, or a possible but unlikely rapid disintegration of part of the Antarctic ice sheet with dramatic consequences for the global sea level. More generally, with a rapidly changing external forcing, the non-linear climate system may experience as yet unenvisionable, unexpected, rapid change.

Observing Anthropogenic Climate Change

Observing the Climate

The question naturally arises whether the system has already undergone human-induced climate change. To answer this question, accurate and detailed observations of climate and climate variability are required. Instrumental observations of land and ocean surface weather variables and sea surface temperature have been made increasingly widely since the mid-19th century. Recently, ships' observations have been supplemented by data from dedicated buoys. The network of upper-air observations, however, only became widespread in the late 1950s. The density of observing stations always has been and still is extremely inhomogeneous, with many stations in densely populated areas and virtually none in huge oceanic areas. In recent times special earth-observation satellites have been launched, providing a wide range of observations of various components of the climate system all over the globe. The correct interpretation of such data still requires high quality in situ and surface data. The longer observational records suffer from changes in instrumentation, measurement techniques, exposure and gaps due to political circumstances or wars. Satellite data also require compensation for orbital and atmospheric transmission effects and for instrumental biases and instabilities. Earlier the problems related to urbanization were mentioned. To be useful for the detection of climate change, observational records have to be adjusted carefully for all these effects.

Concern has been expressed about the present condition of the observational networks. The number of upper-air observations, surface stations and observations from ships is declining, partly compensated for by an increasing number of satellite observations. An increasing number of stations are being automated, which may have an impact on the quality and homogeneity of the observations. Maintaining and improving the quality and density of existing observational networks is essential for necessary high standard information. In order to implement and improve systematic observations of all components of the climate system, the World Meteorological Organization and the International Oceanographic Commission have established a Global Climate Observing System (GCOS). Initially GCOS uses existing atmospheric, oceanic and terrestrial networks. Later GCOS will aim to amplify and improve the observational networks where needed and possible.

Observations alone are not sufficient to produce a coherent and global picture of the state of the climate system. So-called data assimilation systems have been developed, which combine observations and their temporal and spatial statistics with model information to provide a coherent quantitative estimate in space and time of the state of the climate system. Data assimilation also allows the estimation of properties which cannot easily be observed directly but which are linked to the observations through physical laws. Some institutions have recently reanalyzed several decades of data by means of the most recent and most sophisticated version of their data assimilation system, avoiding in this way inhomogeneity due to changes in their system. However in homogeneities in these reanalyzes may still exist due to changing sources of information, such as the introduction of new satellite systems.

The 20th Century

Historically, human activities such as deforestation may have had a local or regional impact, but there is no reason to expect any large human influence on the global climate before the 20th century. Observations of the global climate system during the 20th century are therefore of particular importance. Chapter presents evidence that there has been a mean global warming of 0.4 to 0.8°C of the atmosphere at the surface since the late 19th century. this increase took place in two distinct phases, the first one between 1910 and 1945, and recently since 1976. Recent years have been exceptionally warm, with a larger increase in minimum than in maximum temperatures possibly related, among other factors, to an increase in cloud cover. Surface temperature records indicate that the 1990s are likely to have been the warmest decade of the millennium in the Northern hemisphere, and 1998 is likely to have been the warmest year. For instrumentally recorded history, 1998 has been the warmest year globally. Concomitant with this temperature increase, sea level has risen during the 20th century by 10 to 20 cm and there has been a general retreat of glaciers worldwide, except in a few maritime regions, e.g. Norway and New Zealand.

Regional changes are also apparent. The observed warming has been largest over the mid- and high-latitude continents in winter and spring. Precipitation trends vary considerably geographically and, moreover, data in most of the Southern Hemisphere and over the oceans are scarce. From the data available, it appears that precipitation has increased over land in mid- and high latitudes of the Northern Hemisphere, especially during winter and early spring, and over most Southern Hemisphere land areas. Over the tropical and the Northern Hemisphere subtropical land areas, particularly over the Mediterranean region during winter, conditions have become drier. In contrast, over large parts of the tropical oceans rainfall has increased.

There is considerable variability of the atmospheric circulation at long time-scales. The NAO for example, with its strong influence on the weather and climate of extratropical Eurasia, fluctuates on multi-annual and multi-decadal time-scales, perhaps influenced by varying temperature patterns in the Atlantic Ocean. Since the 1970s the NAO has been in a phase that gives stronger westerly winds in winter. Recent ENSO behaviour seems to have been unusual compared to that of previous decades: there is evidence that El Niño episodes since the mid-1970s have been relatively more frequent than the opposite La Niña episodes.

There are suggestions that the occurrence of extreme weather events has changed in certain areas, but a global pattern is not yet apparent. For example, it is likely that in many regions of the world, both in the Northern and Southern Hemisphere, there has been a disproportionate increase in heavy and extreme precipitation rates in areas where the total precipitation has increased. Across most of the globe there has been a decrease in the frequency of much below-normal seasonal temperatures.

Detection and Attribution

The fact that the global mean temperature has increased since the late 19th century and that other trends have been observed does not necessarily mean that an anthropogenic effect on the climate system has been identified. Climate has always varied on all time-scales, so the observed change may be natural. A more detailed analysis is required to provide evidence of a human impact. Identifying human-induced climate change requires two steps. First it must be demonstrated that an observed climate change is unusual in a statistical sense. This is the detection problem. For this to be successful one has to know quantitatively how climate varies naturally. Although estimates have improved since the SAR, there is still considerable uncertainty in the magnitude of this natural climate variability. The SAR concluded nevertheless, on the basis of careful analyses, that "the observed change in global mean, annually averaged temperature over the last century is unlikely to be due entirely to natural fluctuations of the climate system".

Having detected a climatic change, the most likely cause of that change has to be established. This is the attribution problem. Can one attribute the detected change to human activities, or could it also be due to natural causes? Also attribution is a statistical process. Neither detection nor attribution can ever be "certain", but only probable in a statistical sense. The attribution problem has been addressed by comparing the temporal and spatial patterns of the observed temperature increase with model calculations based on anthropogenic forcing by greenhouse gases and aerosols, on the assumption that these patterns carry a fingerprint of their cause. In this way the SAR found that "there is evidence of an emerging pattern of climate response to forcing by greenhouse gases and sulphate aerosols in the observed climate record".

Evidences of Climate Change

Climate scientists have been constantly engaged in gathering information on key parameters of climate that indicates that some of these parameters have been unequivocally reflect warming of climate. These evidences are gathered from various sources for last hundred years as well as in extensive details for the last two to three decades. Various sources used for gathering such information are innumerable weather stations across the globe, satellites, ships, aircrafts, weather balloons and buoys. These evidences are collated and trend analysis has been undertaken by numerous scientists engaged at national and international level. There are certain parameters that provide indication of warming of climate. Some of these indicators are as follows:

- Increasing temperatures over land and ocean surfaces,
- Melting glacial ice and sea ice,
- Rising sea level,
- Increasing humidity.

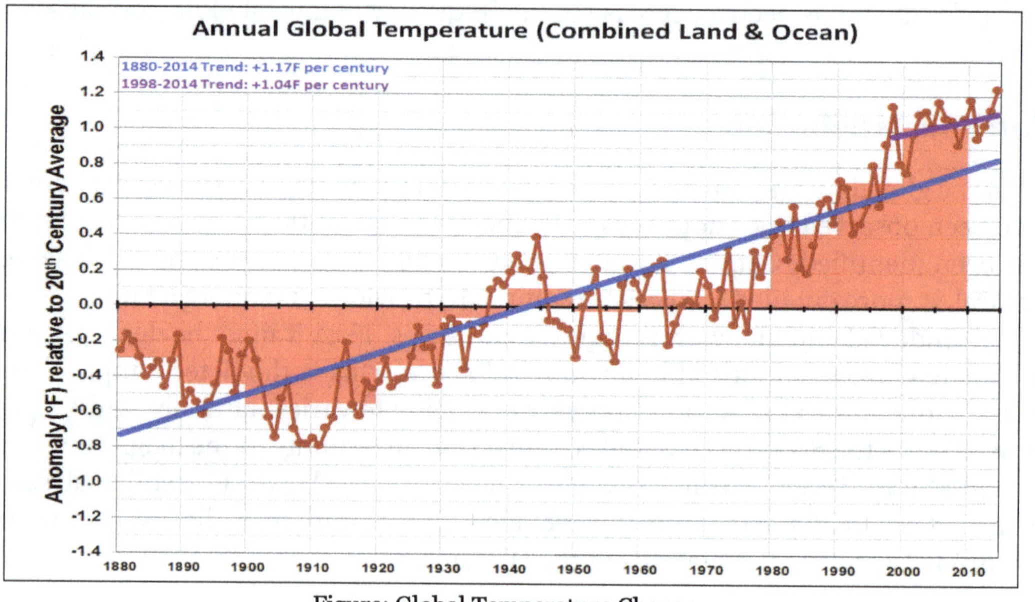

Figure: Global Temperature Change.

Increasing Temperatures Over Land and Ocean Surfaces

If we analyses temperature data in terms of hemisphere since 1880, it has been observed that the years with the warmest land-surface temperatures were 2005 and 2010 in the Northern Hemisphere. In the Southern Hemisphere, 2009 was the warmest. It has also been observed that the period from 2000 to 2010 was the warmest decade since 1880). The data from long-term climate reconstructions of temperature point to the present time as the warmest in the last 120,000 years. These reconstructions also suggest that the increase in temperature during the twentieth century is extremely likely (within a confidence interval of greater than 95%–100 %) the largest to occur in any century over the past 1000 years. Record setting summer daytime temperatures are being recorded in many countries.

According to National Oceanic Administration of America (NOAA) ocean temperatures have also been rising. According to NOAA sea surface temperatures increased at an average rate of 0.07 °C (0.13 °F) per year from 1901 to 2012 as oceans absorbed atmospheric heat. This rise is reflected in measurements of upper-ocean heat content, which includes the upper 700 m (2296 ft.) of ocean.

Melting of Glacial and Sea Ice

As temperatures rise in Earth's atmosphere, glaciers are losing ice mass, shrinking in size. This process is known as glacial retreat. Earth's two largest ice sheets, in Greenland and Antarctica, are also losing ice mass. This is evident from the satellite records. Analysis of satellite data of July 2012 revealed that 97% of the ice sheet's surface was melting. This was the greatest extent in the 30-year record of satellite measurements. Another analysis related to summer melt of the Greenland ice sheet has also revealed that it has increased 30% from 1979 to 2006. On the basis of these evidences, scientists now estimate that between one and two-thirds of Arctic permafrost will thaw over the next 200 years. As you know, these permafrost reserves took thousands of years to form.

Similar kind of observation has also been noticed in relation to sea ice. Maximum sea ice cover in the world is found in the Arctic Sea. The extent of Arctic Sea ice varies over the course of a year. Every summer, some amount of sea ice thaws whereas in winter, the ice refreezes. It has been observed with the help of satellite data that the minimum extent of summer sea-ice occurs in September whereas maximum extent of winter sea-ice occurs in the months of February or early March. However, analysis of satellite data revealed that this has declined since 1979. September sea ice is declining at a rate of 11% per decade in comparison to the 1979– 2000 average and reached its lowest extent in the modern record in 2012. The accelerating decline of summer sea ice, in association with record losses of sea ice in 2007 and 2012, suggests that summer sea ice may disappear sooner than predicted by most models. Some scientists have also estimated an ice-free summer Arctic Ocean within the next few decades.

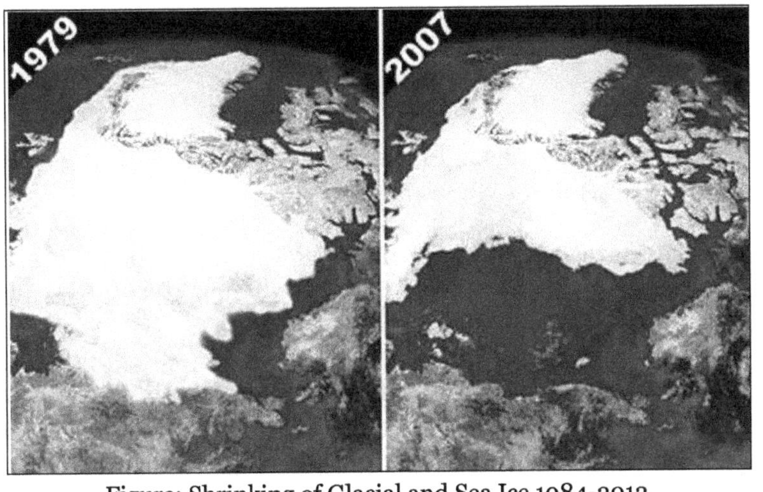

Figure: Shrinking of Glacial and Sea Ice 1984-2012.

Rising Sea-Level

This is another important indicator of climate change. It has been observed that sea level is rising more quickly than the prediction simulated by most of the climate models. During the last century, sea level rose 17–21 cm (6.7–8.3 in.). However, this rise is not uniform across the globe. A greater rise has been observed in some areas like Atlantic coast of U.S. than at any time during the past 2000 years. How did the scientists arrive at such conclusion? Do you have any idea about the tools used for assessing the rise in sea level? Generally we use tidal gauges and satellites to generate such kind of data. Tidal gauge records from 1901 to 2010 show that sea level rose at a rate of 1.7 mm (0.07 in.) per year. Satellite data for the period 1993 - 2013 show that sea level rose 3.16 mm (0.12 in.) per year. This rise is primary due to two major factors that are presently contributing to sea-level rise. About two-thirds of the rise comes from the melting of glaciers and ice sheets whereas the rest one third comes from the thermal expansion of seawater due to absorption of atmospheric heat and expand in volume.

Increasing Humidity and Extreme Events

It has been observed that global average specific humidity has increased by about 0.1 g of water vapour per kilogram of air per decade since 1973. This change is consistent with rising air temperatures, since warm air has a greater capacity to absorb water vapour. A greater amount of water vapour in the atmosphere affects weather in a number of ways and can lead to "extreme" events involving temperature, precipitation, and storm intensity. The decade from 2001 to 2010 showed evidence of a worldwide increase in extreme events, notably heat waves, increased precipitation, and floods. However, to establish a strong linkage between trends related to extreme weather and climate change requires data for a longer timeframe than what is now available.

These above mentioned change have differential impacts in terms of geographical locations, sectors of economy and socio-economic groups. Therefore, some of the major impacts/ consequences of climate change are mentioned for two levels. One is at macro scale i.e. at global level and the other at micro-level i.e. at national level. Some of the major Fifth Assessment Report by IPCC is given below:

Changes in Climate at Global Level

Some of the highlights of the 5th Assessment Report are as follows:

- Each of the last three decades has been successively warmer at the earth's surface than any preceding decade since 1850.
- Ocean warming dominates the increase in energy stored in the climate system, accounting for more than 90% of the energy accumulated between 1971 and 2010. Further uptake of carbon by the ocean will increase ocean acidification.
- Over the last two decades, the Greenland and Antarctic ice sheets have been losing mass, glaciers have continued to shrink almost worldwide, and arctic sea ice and northern Hemisphere spring snow cover have continued to decrease in extent.
- The rate of sea-level rise since the mid-nineteenth century has been larger than the mean rate during the previous two millennia. Over the period 1901–2010, global mean sea level rose by 0.19 m (0.6 ft.).
- The atmospheric concentrations of carbon dioxide (CO_2), methane, and nitrous oxide have increased to levels unprecedented in at least the last 800,000 years, primarily from fossil-fuel emissions and secondarily from net land-use-change emissions. The ocean has absorbed about 30% of the emitted anthropogenic carbon dioxide, causing ocean acidification.
- Total radiative forcing is positive and has led to an uptake of energy by the climate system. The largest contribution to total radiative forcing is caused by the increase in the atmospheric concentration of CO_2 since 1750.

- Warming of the climate system is unequivocal. Many of the temperature changes observed since the 1950s are unprecedented over decades to millennia. It is extremely likely (95%–99%) that human influence has been the dominant cause of the observed warming since the mid-twentieth century.
- Climate models have improved since the Fourth Assessment Report. Models reproduce observed continental-scale surface temperature patterns and trends over many decades, including the more rapid warming since the mid-twentieth century and the cooling immediately following large volcanic eruptions.
- Continued emissions of greenhouse gases will cause further warming and changes in all components of the climate system. Limiting climate change will require substantial and sustained reductions of greenhouse gas emissions.
- Changes in the global water cycle will not be uniform. The contrast in precipitation between wet and dry regions and between wet and dry seasons will increase.
- Global mean sea level will continue to rise. The rate of sea-level rise will very likely exceed that observed during 1971–2010, due to increased ocean warming and increased loss of ice mass from glaciers and ice sheets.

Climate Change Mitigation

Climate Change Mitigation refers to efforts to reduce or prevent emission of greenhouse gases. Mitigation can mean using new technologies and renewable energies, making older equipment more energy efficient, or changing management practices or consumer behavior. It can be as complex as a plan for a new city or as a simple as improvements to a cook stove design. Efforts underway around the world range from high-tech subway systems to bicycling paths and walkways.

There are three main climate change mitigation approaches.

First, conventional mitigation efforts employ decarburization technologies and techniques that reduce CO_2 emissions, such as renewable energy, fuel switching, efficiency gains, nuclear power, and carbon capture storage and utilization. Most of these technologies are well established and carry an acceptable level of managed risk.

A second route constitutes a new set of technologies and methods that have been recently proposed. These techniques are potentially deployed to capture and sequester CO_2 from the atmosphere and are termed negative emissions technologies, also referred to as carbon dioxide removal methods. The main negative emissions techniques widely include bioenergy carbon capture and storage, bio char, enhanced weathering, direct air carbon capture and storage, ocean fertilization, ocean alkalinity enhancement, soil carbon sequestration, afforestation and reforestation, wetland construction and restoration, as well as alternative negative emissions utilization and storage methods such as mineral carbonation and using biomass in construction.

Finally, a third route revolves around the principle of altering the earth's radiation balance through the management of solar and terrestrial radiation. Such techniques are termed radioactive forcing reengineering technologies, and the main objective is temperature stabilization or reduction. Unlike negative emissions technologies, this is achieved without altering greenhouse gas concentrations in the atmosphere. The main radiative forcing reengineering techniques that are include stratospheric aerosol injection, marine sky brightening, cirrus cloud thinning, space-based mirrors, surface-based brightening and various radiation management techniques. All these techniques are still theoretical or at very early trial stages and carry a lot of uncertainty and risk in terms of practical large-scale deployment. At the moment, radiative forcing geo engineering techniques are not included within policy frameworks.

Conventional Mitigation Technologies

Energy-related emissions are the main driver behind the increased greenhouse gas concentration levels in the atmosphere; hence, conventional mitigation technologies and efforts should be focused on both the supply and demand sides of energy. Mitigation efforts primarily cover technologies and techniques that are deployed in four main sectors, power on the supply side and industry, transportation and buildings on the demand side. Within the power sector, decarbonization can be achieved through the introduction of renewable energy, nuclear power, carbon capture and storage as well as supply-side fuel switches to low-carbon fuels such as natural gas and renewable fuels. Furthermore, mitigation efforts on the demand side include the efficiency gains achieved through the deployment of energy-efficient processes and sector-specific technologies that reduce energy consumption, as well as end-use fuel switch from fossil-based fuels to renewable fuels, and, moreover, the integration of renewable power technologies within the energy matrix of such sectors.

Carbon Capture, Storage and Utilization

Carbon capture and storage is a promising technology as a potential decarbonization approach to be applied to the power as well as the industrial sectors. The technology consists of separating and capturing CO_2 gases from processes that rely on fossil fuels such as coal, oil or gas. The captured CO_2 is then transported and stored in geological reservoirs for very long periods. The main objective is the reduction in emission levels while utilizing fossil sources. Three capturing technologies are: pre-combustion, post-combustion and ox fuel combustion. Each technology carries a specific process to extracting and capturing CO_2. Post-combustion capture technologies, however, are the most suitable for retrofit projects and have vast application potential. Once CO_2 has been successfully captured, it is liquefied and transported through pipelines or ships to suitable storage sites. Storage options include depleted oil and gas Felds, coal beds and underground saline aquifers not used for potable water. Some of the main drawbacks of carbon capture and storage include safety in relation to secured storage and the possibility of leakage. Negative environmental impacts that may result from onshore storage locations that undergo accidental leakage have been investigated. The investigation focused on the impact of leakage on agricultural land. Risk of leakage and associated negative impacts have also been pointed out. Other issues related to this technology include public acceptance as well as the high deployment costs associated). Another pathway post carbon capture is the utilization of the CO_2 captured in the production of chemicals, fuels, microalgae and concrete building materials, as well as utilization in enhanced oil recovery.

Large-scale deployment of carbon capture storage and utilization technologies is yet to be proven. There are 9 more carbon capture projects under development and are projected to increase capacity to 11 $MtCO_2$ by 2025; however, a significant deviation exists from the sustainable development scenario targeted by the international energy agency for 2040 which is a capacity of 1488 $MtCO_2$.

Fuel Switch and Efficiency Gains

Fuel switching in the power sector from coal to gas, in the short-term, has been a potential approach to economically transition to a low carbon and hopefully a zero-carbon economy in future. The move to natural gas is also applicable to industry, transportation and building sectors; however, the switch to renewable fuels is a more sustainable approach creating further decarburization potential in these sectors.

In addition to fuel switching, efficiency gains are of extreme significance within mitigation efforts. Efficiency gains in the power sector are achieved through improvements in thermal power plants by enhancing the efficiency of fuel combustion as well as improving turbine generator efficiencies. Furthermore, waste heat recovery for additional thermal as well as electric production enhances efficiency. In gas-Fred power plants, the utilization of a combined cycle technology enhances the efficiency significantly. Combined heat and power units have also played an interesting role in efficiency gains. Technological advances within transmission and distribution networks also enhance efficiencies by reducing losses.

In industry, there are many potential areas where efficiency gains may be realized. For example, in steel and cement applications, waste heat can be recovered for onsite power and heat production through the installation of waste heat-driven power plants that utilize waste heat from exhaust gases. For industries that utilize process steam, there is an excellent opportunity to utilize waste steam pressure to generate electric power for onsite usage or drive rotating equipment. The application of back pressure steam turbines in areas where steam pressure reduction is required can enhance energy efficiency significantly. The same approach can be deployed in applications where gas pressure reduction is required, however, using turbo expanders. Waste gases from industrial processes can also be utilized to generate onsite heat and power using micro- and small gas turbines. In addition, further efficiency gains can be realized through the deployment of advanced machinery controls in a multitude of processes and industrial sectors.

A number of factors influence energy efficiency within buildings, first the building design as well as materials utilized in construction, e.g. insulation and glazing. Furthermore, appliances, devices and systems used throughout buildings, e.g. heating, cooling and ventilation systems, and lighting, play a pivotal role in energy consumption. Efficiency gains can be realized by utilizing energy-efficient systems and appliances as well as improved construction materials.

In the transportation sector, efficiency gains can be realized through the introduction of enhanced and more efficient thermal engines, hybrid and electric vehicles as well as hydrogen (H_2) vehicles. Furthermore, efficiency gains can be achieved through technological advances within aviation, shipping and rail, although rail is currently one of the most energy-efficient modes. Efficiency measures in the transportation sector can also take other forms. For example, travel demand management, to reduce frequency and distance of travel, can be an interesting approach.

Moreover, shifting travel to the most efficient modes where possible, such as electrified rail, and reducing dependence on high-intensity travel methods can play an interesting role in enhancing efficiency.

Negative Emissions Technologies

Most of the climate pathways that were investigated by the Intergovernmental Panel on Climate Change (IPCC) included the deployment of negative emissions technologies along with conventional de carbonization technologies to assess the feasibility of achieving the targets mandated by the Paris agreement. Only two negative emissions technologies have been included in the IPCC assessments so far, bioenergy carbon captures and storage as well as afforestation and reforestation (IPCC 2018).

Empirically investigated the potential negative emissions needed to limit global warming to less than 2 °C. The analysis utilized an IPCC pathway that is most likely to maintain warming at such level and constructed a number of scenarios based on conventional mitigation assumptions in an attempt to quantify the potential negative emissions efforts required. The results indicated that in the best-case scenario that is under the best assumptions on conventional mitigation efforts negative emissions of 0.5–3Gt C year−1 and 50–250 Gt C of storage capacity are required. Based on a worst-case scenario, negative emissions of 7–11 Gt C year−1 and 1000–1600 Gt C of storage capacity are required. (1 Gigaton Carbon=3.6667 Gigaton CO_2e) The results indicate the inevitable need for negative emissions, even at very high rates of conventional mitigation efforts. Furthermore, the study suggests that negative emissions alone should not be relied upon to meet the 2 °C target. The investigation concluded that since negative emissions technologies are still at an infant stage of development, conventional mitigation technologies should remain focused upon within climate policy, while further financial resources are to be mobilized to accelerate the development of negative emissions technologies.

It is argued that negative emissions technologies should be deployed to remove residual emissions after all conventional De carbonization efforts have been maximized and that such approach should be utilized to remove emissions that are difficult to eliminate through conventional methods. It is important to note that negative emissions should be viewed as a complementary suite of technologies and techniques to conventional decarbonization methods, and not a substitute.

The significant role of negative emissions in meeting climate targets is understood and appreciated amongst academics, scientists and policymakers; however, there still remains a debate on the social, economic and technical feasibility as well as the risk associated with large-scale deployment. Here we will carry out the main negative emissions technologies and techniques, their current state of development, perceived limitations and risks as well as social and policy implications.

Bioenergy Carbon Capture and Storage

Bioenergy carbon capture and storage, also referred to as BECCS, is one of the prominent negative emissions technologies discussed widely in the literature. The Intergovernmental Panel on Climate Change (IPCC) heavily relied on bioenergy carbon capture and storage within their assessments as a potential route to meet temperature goals (IPCC 2018). The technology is simply an integration of bio power, and carbon capture and storage technologies discussed earlier. The basic principle behind the technology is quite straightforward. Biomass biologically captures atmospheric CO_2 through photosynthesis during growth, which is then utilized for energy production through combustion. The CO_2 emissions realized upon combustion are then captured and stored in suitable geological reservoirs). This technology can significantly reduce greenhouse gas concentration levels by removing CO_2 from the atmosphere. The carbon dioxide removal potential of this technology varies within the literature; however, a conservative assessment by Fuss et al. presents an estimated range of 0.5–5 $GtCO_2$ year−1 by 2050. In terms of global estimates for storage capacity, the literature presents a wide range from 200 to 50,000 $GtCO_2$. Cost estimates for carbon dioxide removal through bioenergy carbon capture and storage are in the range of \$100-\$200/tCO_2.

The biomass feedstock's utilized for this approach can either be dedicated energy crops or wastes from agricultural or forestry sources. Furthermore, such feedstock can either be used as dedicated bio-based feedstock's or can be combined with fossil-based fuels in co-fred power plants. Besides the standard combustion route, the literature suggests that CO_2 can be captured in non-power bio-based applications, such as during the fermentation process in ethanol production or the gasification of wood pulp effluent, e.g. black liquor, in pulp production.

The main challenge associated with this technology is the significant amount of biomass feedstock required to be an effective emission abatement approach. Under large-scale deployment, resource demand when utilizing dedicated crops would be quite significant, with high pressure exerted on land, water as well as nutrient resources. A major issue would be the direct competition with food and feed crops for land, freshwater and nutrients. Heck et al. empirically investigated the large-scale deployment of bioenergy carbon capture and storage for climate change abatement and demonstrated its impact on freshwater use, land system change, biosphere integrity and biogeochemical fowl. Furthermore, the investigation identified the interrelated nature between each of these dimensions as well as the associated impacts when any one dimension is prioritized. A sustainable approach to land use is quite critical in approaching bioenergy carbon capture and storage. Competing with food for arable land and changing forest land to dedicated plantations has serious negative social and environmental effects. The effectiveness of this technology in achieving negative emissions is based on several factors which include previous land cover, the initial carbon gain or loss due to land-use change, bioenergy crop yields, and the amount of harvested carbon that is ultimately sequestered. Their empirical investigation highlights the negative impact of bioenergy carbon capture and storage when dedicated plantations replace carbon-dense ecosystems. Another issue is the albedo effects of biomass cultivation. This is mainly applicable in high-latitude locations, where biomass replaces snow cover and reduces radiation refection potential which offsets mitigation efforts.

In terms of technology readiness, bioenergy technologies are to a certain extent well developed; however, carbon capture and storage are still at an early stage. Technology risk is mainly associated with storage integrity and the potential of leakage on carbon capture and storage. Besides, they question whether this technology can deliver its abatement potential within the projected time frame. In terms of policy, it is argued that a strong framework, as well as adequate incentives, needs to be in place to properly push the technology forward. Commercial logic may not be enough to drive forward global deployment. Financial viability of such projects will depend on a utilitarian carbon market that caters for negative emissions as well as an appropriate carbon price that incentivizes deployment. Therefore, policy should look at ways to strengthen carbon pricing mechanisms and introduce negative emissions as a new class of tradable credits.

Afforestation and Reforestation

During tree growth, CO_2 is captured from the atmosphere and stored in living biomass, dead organic matter and soils. Forestation is thus a biogenic negative emissions technology that plays an important role within climate change abatement efforts. Forestation can be deployed by either establishing new forests, referred to as afforestation, or re-establishing previous forest areas that have undergone deforestation or degradation, which is referred to as reforestation. Depending on tree species, once forests are established CO_2 uptake may span 20–100 years until trees reach maturity and then sequestration rates slow down significantly. At that stage, forest products can be harvested and utilized. It is argued that forest management activities and practices have an environmental impact and should be carefully planned. Several advantages and co-benefits those are associated with forest-based mitigation which include biodiversity, food control as well as quality improvement for soil, water and air.

Carbon can be stored in forests for a very long time; however, permanence is vulnerable due to natural and human disturbances. Natural disasters such as free, droughts and disease or human-induced deforestation activities are all risks that negatively impact storage integrity. In general, biogenic storage has a much shorter lifespan than storage in geological formations, such as in the case of bioenergy carbon capture and storage. Another issue related to forestation is land requirement as well as competition with other land use. Significant amounts of land are required to achieve effective abatement results. Forests in high latitudes would actually be counterproductive, accelerating local warming as well as ice and snow cover loss. They argue that tropical areas would be the most suitable zones to host forestation projects. However, competition with agriculture and other sectors for land will be another problem. Based on global tropical boundary limitations, an estimated total area of 500 Mha is argued to be suitable for forestation deployment. This would allow for a global carbon dioxide removal potential of 0.5–3.6 $GtCO_2$ year–1 by 2050. Removal costs are estimated at \$5–\$50/tCO_2.

In terms of technology readiness, afforestation and reforestation have already been widely adopted on a global level and have already been integrated within climate policies through the Kyoto protocol's clean development mechanism program. To drive forward forest based mitigation efforts, the protocol introduced removal units which allowed forestation projects to yield tradable credits. Despite the early policy measures, forest-based mitigation efforts accounted for a small fraction of emissions at that time. Forest-based abatement projects have also been introduced through national regulations as well as voluntary systems such as the reducing emissions from deforestation and forest degradation (REDD+) programme that was introduced by the United Nations in 2008. However, carbon sequestration through forestation remained insignificant, as it only accounted for 0.5% of the total carbon traded in 2013.

The effectiveness of the REDD+ programme is argued in the literature after more than 10 years of its introduction. Hein et al. present a number of arguments around the program's poor track record in achieving its intended purpose of emissions reduction. However, despite the uncertainty and weaknesses discussed, REDD+ implementation intentions have been indicated by 56 countries in their INDC submissions under the Paris agreement. Permanence, sequestration uncertainty, the availability of efficient financing mechanisms as well as monitoring, reporting and verification systems are all difficulties associated around forest-based abatement projects.

Biochar

Biochar has recently gained considerable recognition as a viable approach for carbon capture and permanent storage and is considered as one of the promising negative emissions technologies. Biochar is produced from biomass, e.g. dedicated crops, agricultural residues and forestry residues, through a thermochemical conversion process. It is produced through pyrolysis, a process of heating in the absence of oxygen, as well as through gasification and hydrothermal carbonization. The carbon captured by biomass through CO_2 uptake during plant growth is then processed into a char that can be applied to soils for extended periods. The conversion process stores biomass carbon in a form that is very stable and resistant to decomposition. Stability in soils is perhaps the most important property of biochar that makes it a solid carbon removal technology. Although considered more stable than soil organic carbon, there are certain uncertainties around decomposition rates of various types of biochar, which depends on the feedstock used and process conditions utilized. Depending on the feedstock used, it is estimated that this technology can potentially remove between 2.1 and 4.8 tCO_2/tone of biochar. Carbon removal potential, as well as costs, however, a conservative range is provided. It is estimated that by 2050 global carbon reduction removal potential achieved through biochar can be in the range of 0.3–2 Gt CO_2 $year^{-1}$, with costs ranging from \$90 to \$120/tCO_2. In terms of resource requirements, biochar production would require vast amounts of land to have an effective impact on greenhouse gas concentration levels. Land is required for feedstock cultivation, as well as for biochar dispersal acting as a carbon sink. While land for dedicated biomass cultivation may create competition issues with agriculture and other land-use sectors, same as the case of bioenergy carbon capture and storage, there would be no issues with areas required for biochar dispersal. This would be the case as long as the biochar is technically matched with the type of crop, soil and growing conditions related to the specific cropping system. Besides soil, other carbon sink applications for biochar which include construction materials, wastewater treatment and electronics, as long as the product does not thermally degrade or oxidize throughout its life cycle. Furthermore, it has been argued in the literature that marginal and degraded lands can potentially be utilized for dedicated plantations, relieving pressure on land that can be used for other purposes. Moreover, using waste biomass eliminates the need for land and provides a waste disposal solution; however, competition over waste for other purposes increases feedstock availability risk as well as price volatility. Biomass availability is one of the limiting factors to successful large-scale deployment of Biochar projects.

In addition to the beneficial effect of capturing and storing CO_2 from the atmosphere, there is growing evidence in the literature that biochar also has an impact on other greenhouse gas emissions such as CH_4 and N_2O. This is specific to the cropping system as well as the type of biochar utilized and its processing conditions. Xiao et al. also present conflicting results regarding biochar application, which is very specific to the condition of the soils amended with biochar. Impact on greenhouse gas emissions should, therefore, be studied on a case-by-case basis.

Another benefit that is the positive effects associated with biochar application to soils. It is argued that soil quality and fertility are significantly enhanced. Improvement in nutrient cycling, reduction in nutrient leaching from the soil and an increase in water and nutrient retention as well as stimulation of soil microbial activity are all co-benefits associated with biochar application. However, this is mainly dependent on biochar physical and chemical properties. Such properties are defined by the type of feedstock utilized, pyrolysis conditions, as well as other processing conditions. Furthermore, despite the general perception that biochar positively impacts plant growth and production, which is true in a large number of cases, there is evidence that biochar application may hinder plant growth in certain cropping systems. This is based on the type of biochar, the quantity applied and the specific crops under cultivation and sometimes management practices. The evidence is mixed, and therefore careful analysis should be carried out to successfully match biochar with appropriate carbon sinks.

Concerning the risks associated with large-scale deployment, with high application rates of biochar to the soil surface, e.g. 30–60 tons/ha, it is argued that a decrease in surface reflectivity would increase soil temperature, which in turn would reduce the beneficial effect of carbon sequestration through this route. Other risks and challenges associated include the risk of reversibility and challenges in monitoring, reporting and verification. Moreover, limited policy incentives and support, as well as lack of carbon pricing mechanisms that incorporate CO_2 removal through biochar, hinder this technology's potential for large-scale commercialization. The role of government policy in accelerating biochar adoption and identified three types of existing policy instruments that can be used to stimulate biochar deployment in the USA: commercial financial incentives, non-financial incentives and research and development funding. With the current technological advancements, in particular block chain, a number of start-ups are developing carbon removal platforms to drive forward voluntary carbon offsets for consumers and corporations. A Finnish start-up, has introduced biochar as a net-negative technology. Once verified through the company's verification system, the carbon removal certificates generated by biochar producers are auctioned to potential offset parties. However, until carbon removal is adequately monetized and supported through sufficient policy instruments, biochar project development will probably not reach the scale required to have a profound impact within the time frame mandated by international policy.

Climate Change and Food Security

Food security exists when all people at all times have physical or economic access to sufficient safe and nutritious food to meet their dietary needs. Global food security refers to the situation where each person, member of any household is having physical and economic access to sufficient, safe and nutritious food to meet their dietary needs and food preferences for an active and healthy life – (Food and Agriculture Organization definition). That food has to be affordable, safe and healthy, culturally acceptable, meeting specific dietary needs of the people and is obtained in a dignified manner, produced in ways that are environmentally sound and socially just. Thus there should be no perceived inadequacy of the household food supply expressed through concerns about running out of food and not having enough food to make a meal or malnourished as a result of physical unavailability of food, lack of socio-economic access to food or inadequate food utilization.

The World's efforts to meet the Millennium Development Goal of cutting hunger in half by 2015 are far from reach. With the world's population set to reach 9 billion by 2050, agricultural production will need to increase by 70% in order to meet demand. Climate change adds a new dimension of this challenge as it is one of the key drivers of change affecting the food system and contributing to rising food prices. It leads to changes in growing seasons and rainfall patterns and the increased frequency of extreme events such as droughts and floods. It has been estimated by the United Nations Environment Programme that up to 25 per cent of world food production could be lost by 2050 as a result of climate change, water scarcity and land degradation.

In the developing world, nearly 1 billion people are unable to meet their dietary needs. Another 5-10% is at risk of 'acute' food insecurity in times of crisis. Despite improvements, the millennium development goal on hunger is likely to be missed by a wide margin in areas like Sub Saharan Africa, where persistent food insecurity is compounded by widespread political instability, conflict and the HIV/AIDS pandemic. The UN Food and Agriculture Organization estimate that 820 million people in developing countries are suffering from malnutrition. In a world in which nearly half the population survives on $2/day or less, more than 800 million people go to sleep hungry any given night, and a child dies every five seconds due to hunger related complications, the need to respond to the needs of the poor for food is ever-present and widespread. Despite the universal recognition of every person's right to food, vulnerability to hunger remains a daily reality for many.

The United States of America and European Union together provide about two thirds of global food aid deliveries. The United States of America is by far the most important donor of food aid both for bilateral programme aid and as the main contributor to the World Food Program. The World Food Program (WFP) is the primary agency responsible for administering multi- lateral food aid. The WFP and various NGOs administer project food aid to support a wide range of developmental projects targeting the poor in developing countries. Others include the UN Food and Agriculture Organization (FAO), the Food Aid Convention (FAC) and the World Trade Organization (WTO). All these organizations have different mandates and are concerned with different aspects of the provision of food aid. Many organizations around the world are working to find ways to produce the food needed in a sustainable way, within the limits of what our ecosystems can support for current and future generations, and to safeguard this production from the impacts of climate change. Food security is the outcome of food system processes all along the food chain. Climate change will affect food security through its impacts on all components of global, national and local food systems. It is important to note that climate change variables influence biophysical factors, such as plant and animal growth, water cycles, biodiversity and nutrient cycling, and the ways in which these are managed. Agriculture is the primary source of livelihood among many rural populations.

Causes of Global Food Insecurity

There are many causes of global food insecurity and most of them are region specific, except climate change where impacts are felt globally. Some are human causes like destruction of fertile lands and others are non-human like natural disasters (e.g. floods). While human destruction of his mother nature through careless management practices have impacted negatively on his ability to produce enough food for the population, the contribution that humans make to climate change is potentially causing increased food insecurity globally. Industrialization, modern agricultural practices and the need to produce for the market without thinking about environmental sustainability has had disastrous effect as climate change has set in. In some areas, sea levels have risen and increased incidence of extreme events posing danger and threatening livelihoods and at the same time increasing the vulnerability to future food insecurity in the world. Coastal freshwater are being contaminated with salt water and people do not get fresh water for domestic use and plants or food crops that are not salt-tolerant cannot be grown. Storm surges become common occurrences together with flooding. This hinders people living along coast from growing food crops. Increased temperature also leads to heat stress for plants which increase evaporation and lowers productivity. Climate change has also been a cause of more frequent and more intense extreme weather events like increasing irregularities in seasonal rainfall patterns. The changing growing seasons have shifted ecological niches.

Rainfall is becoming more unpredictable and unreliable and has become a common occurrence leading to greater uncertainty among farmers and their traditional agricultural knowledge and coping strategies. The phenomenon has had immediate impact on food production, distribution, infrastructure, livelihoods and human health in all parts of the world. Where climate change has influenced rainfall and temperature patterns, the suitability of land for different types of crops and pasture is affected, including the health and productivity of forests; the distribution, productivity and community composition of marine resources and the increased incidence of pests and diseases. It also affects the functioning of biodiversity and ecosystem of natural habitats; and the availability of good-quality water for crops, livestock and inland fish production. It may also increase aridity of arable lands, induce drought and deforestation, can increase fire danger with consequent loss of the vegetative cover needed for grazing and fuel wood. It also leads to depletion groundwater and induce the internal and international migration thereby triggering resource-based conflicts and civil unrest in either areas of origin or destination. Conflicts over water resources will have implications for both food production and people's access to food in conflict zones. Production from both rain-fed and irrigated agriculture in dry land ecosystems accounts for approximately 25 percent, and rice produced in coastal ecosystems for about 12 percent.

Natural fall-back mechanisms during food crisis may also be affected. These include disappearance of traditional fruits, herbs, vegetables, mushrooms, wild foods and other coping mechanisms. It is expected that as the world climate continues to change, 5 000 plant species in sub-Saharan Africa will decrease in size or shift to other agro-ecological zones due to climate change. All these impacts have negative influence on food security of the people living in the affected areas. At the same time people may resort to unfamiliar ways of surviving and countering food insecurity. It was observed that people may decide not to migrate but find new, unfamiliar ways of earning a living. FAO has been instrumental in assisting various communities all over the World develop the right food production technologies aimed at boosting food security. They possesses technical expertise relevant to climate change adaptation in a variety of ecosystems, including agro-ecosystems (crops, livestock, grasslands), forests and woodlands, inland waters, and coastal and marine ecosystems. It works to build national, local and community-level capacities to raise awareness of and prepare for climate change impacts, assists member countries in identifying potential adaptation options and helps local people understand which are the most applicable to their particular circumstances.

Since 2002, FAO has been promoting National and Regional Programmed for Food Security (NPFS and RPFS) as instruments that help countries enhance productivity and diversify the livelihoods of rural people on a scale sufficient to achieve the 2015 targets set by WFS and the Millennium Development Goals (MDGs). This is an organization committed to eliminating food insecurity in the World. From their advice, people have created artificial microclimates, breed plants and animals with desired characteristics, enhance soil quality, and control the flow of water. Advances in storage, preservation and transport technologies have made food processing and packaging a new area of economic activity. This has allowed food distributors and retailers to develop long-distance marketing chains that move produce and packaged food. The countries that are not able to adopt new technologies need to be assisted. Developing countries urgently need more assistance to help them prepare for the impacts of climate change that are unavoidable. The transfer of the industrialized countries' best energy efficiency and renewable energy technology and assistance with disaster preparedness, agricultural productivity improvements, water management, conflict prevention, reforestation, preventing deforestation and critical infrastructure would be most appropriate.

Other important practices for addressing food insecurity include changing consumption patterns and food preparation practices, efficient water use, improving soil quality, capacity to withstand extreme events and carbon sequestration. Others include promotion of agro biodiversity for local adaptation and resilience, reducing uncertainty by improving the information base and devising innovative schemes for insuring against climate change hazards.

People need to adopt practices that enable the vulnerable to protect their existing livelihood systems, diversify their sources of income, change their livelihood strategies or migrate. Sustainable livestock management practices for adaptation and associated mitigation should also be given high priority.

Integrating Indigenous Knowledge and Science based Technologies

Integration of both indigenous knowledge and science based technologies is important in combating the effects of climate change on food security. Farmers need to combine the best of their traditional approaches with modern agro-science based technology. Towards this end documenting traditional knowledge of the world is paramount if an integrated approach is to be effective as most traditional knowledge remains as tacit knowledge.

Climate Change, Gender and Human Health

Effects of climate change on health will impact on most populations in the coming decades and put the lives and well-being of billions of people at increased risk. IPCC states that "climate change is projected to increase threats to human health".

Climate change can affect human health through a range of mechanisms. These include relatively direct effects of hazards such as heat waves, floods and storms, and more complex pathways of altered infectious disease patterns, disruptions of agricultural and other supportive ecosystems, and potentially population displacement and conflict over depleted resources, such as water, fertile land and fisheries. There is no clear dividing line between these divisions, and each pathway is also modulated by non-climatic determinants and human actions.

Health, Gender and Climate Change

Limited case examples and research have analysed and highlighted the links between gender norms, roles, relations and health impacts of climate change (Box). The framework in figure, adapted from the synthesis report of the International Scientific Congress on Climate Change, is used in this paper to structure the available information on the gendered health implications of climate change, according to (i) the direct and indirect health impacts of meteorological conditions; (ii) the health implications of potential societal effects of climate change, for example on livelihoods, agriculture and migration; and (iii) capacities, resources, behaviours and attitudes related to health adaptation measures and mitigation policies that have health implications.

Why Gender and Health?

The distinct roles and relations of men and women in a given culture, dictated by that culture's gender norms and values, give rise to gender differences. Gender norms, roles and relations also give rise to gender inequalities – that is, differences between men and women that systematically value one group often to the detriment of the other. The fact that, throughout the world, women on average have lower cash incomes than men is an example of gender inequality. Both gender differences and gender inequalities can give rise to inequities between men and women in health status and access to health care. For example:

- A woman cannot receive needed health care because norms in her community prevent her from travelling alone to a clinic,
- An adolescent boy dies in an accident because of trying to live up to his peers' expectations that young men should be "bold" risk-takers, including on the road.

Impact pathways		**Current responses**	
Meteorological conditions exposure	**Human/social consequences of climate change**	**Mitigation actions**	**Adaptation actions**
Examples: • Warming • Humidity • Rainfall/drying • Winds • Extreme events	*Examples:* • Displacement • Shift in farming and land use	*Examples:* • Alternative energy • Accessible clean water	*Examples:* • Addressing water shortage • Crop substitution • Community education on early warning systems and hazard management
Examples of impact outcomes and responses that are gendered in their effects			
• Injury/death from hunger • Epidemics • Mental health issues • Water-related infections	• Migration • Exacerbation of malnutrition • Increased violence against women and girls	• Hydropower – leading to more snail hosts for schistosomiasis • Cleaner air – less cardiorespiratory diseases (gendered profiles)	• Unexpected nutrient deficiencies • Impacts of water quality • Fewer deaths in extreme events

Figure: Effects of climate change on human health and current responses: a gendered perspective.

In each of these cases, gender norms and values, and resulting behaviours, are negatively affecting health. But gender norms and values are not fixed and can evolve over time, can vary substantially from place to place, and are subject to change. Thus, the adverse health consequences resulting from gender differences and gender inequalities are not static. They can be changed.

Impacts Health

Meteorological Conditions and Human Exposure

There is good evidence showing that women and men suffer different negative health consequences following extreme events such as floods, windstorms, droughts and heat waves. A review of census information on the effects of natural disasters across 141 countries showed that although disasters create hardships for everyone, on average they kill more women than men, or kill women at a younger age than men. These differences persist in proportion to the severity of disasters and depend on the relative socioeconomic status of women in the affected country. This effect is strongest in countries where women have very low social, economic and political status. In countries where women have comparable status to men, natural disasters affect men and women almost equally (same study highlighted that physical differences between men and women are unlikely to explain these differences, and social norms may provide some additional explanation. The study also looked at the specific vulnerability of girls and women with respect to mortality from natural disasters and their aftermath; the study found that natural disasters lower the life expectancy in women more than in men. Since life expectancy of women is generally higher than that of men, natural disasters actually narrow the gender gap in life expectancy in most countries. The research also confirmed that the effect on the gender gap in life expectancy is proportional to the severity of disasters – that is, major calamities lead to more severe impacts on women's life expectancy compared with that of men. The study verified that the effect of the gender gap on the gender gap in life expectancy varied inversely in relation to women's socioeconomic status. This highlights the socially constructed and gender-specific vulnerability of women to natural disasters, which is integral to everyday socioeconomic patterns and leads to relatively higher disaster-related mortality rates in women compared with men.

Heat Waves and Increased Hot Weather

Warming and increased humidity have already contributed to observed increases in some health risks, and these can be anticipated to continue in the future.

Direct Consequences

Several studies, mainly in cities in developed countries, have shown that death rates increase as temperatures depart, in either direction, from the optimum temperature for that population. There is therefore concern that although warmer temperatures may lead to fewer deaths in winter, they are likely to increase summer mortality. For example, it is estimated that a 2 °C rise would increase the annual death rate from heat waves in many cities by approximately twofold. There is evidence that vulnerability varies by sex: more women than men died during the 2003 European heat wave, and the majority of European studies have shown that women are more at risk, in both relative and absolute terms, of dying in such events. There may be some physiological reasons for an increased risk among elderly women. Social factors can also be important in determining the risk of negative impacts of heat waves. For example, in the United States of America, elderly men seem to be more at risk than women in heat waves, and this was particularly apparent in the Chicago events of July 1995. This vulnerability may be due to the level of social isolation among elderly men. In Paris, France the heat wave-related risk increased for unmarried men but not for unmarried women. Men may also be more at risk of heatstroke mortality because they are more likely than women to be active in hot weather.

Indirect Consequences

Rising temperatures may increase the transmission of malaria in some locations, which already causes 300 million acute illnesses and kills almost 1 million people every year. Pregnant women are particularly vulnerable to malaria as they are twice as "appealing" as non-pregnant women to malaria-carrying mosquitoes. A study that compared the relative "attractiveness" to mosquitoes of pregnant and non-pregnant women in rural Gambia found that the mechanisms underlying this vulnerability during pregnancy is likely to be related to at least two physiological factors. First, women in the advanced stages of pregnancy (mean gestational age 28 weeks or above) produce more exhaled breath (on average, 21% more volume) than their non-pregnant counterparts. There are several hundred different components in human breath, some of which help mosquitoes detect a host. At close range, body warmth, moist convection currents, host odours and visual stimuli allow the insect to locate its target. During pregnancy, blood flow to the skin increases, which helps heat dissipation, particularly in the hands and feet. The study also found that the abdomen of pregnant women was on average 0.7 °C hotter than that of non-pregnant women and that there may be an increase in the release of volatile substances from the skin surface and a larger host signature that allows mosquitoes to detect them more readily at close range. Changes in behaviour in pregnant women can also increase exposure to night-biting mosquitoes: pregnant women leave the protection of their bed net at night to urinate twice as frequently as non-pregnant women. Although the important role of immunity and nutrition is recognized, it is suggested that physiological and behavioural changes that occur during pregnancy could partly explain this increased risk of infection. Maternal malaria increases the risk of spontaneous abortion, premature delivery, stillbirth and low birth weight.

Evidence for connections between weather and preeclampsia varies between studies. Some studies have looked at links between meteorological conditions and the incidence of eclampsia in pregnancy; the studies found increased incidence during climatic conditions characterized by low temperature, high humidity or high precipitation, with an increased incidence especially during the first few months of the rainy season. A study from Kuwait found that incidence of preeclampsia was high in November, when the temperature was low and the humidity high. On the other hand, the incidence of pregnancy-induced hypertension was highest in June, when the temperature was very high and the humidity at its lowest. Another study, from the southern province of Zimbabwe, evaluated hypertensive complications during pregnancy and observed a distinctive change in the incidence of preeclampsia during the year. These changes corresponded with the seasonal variation in precipitation, with incidence increasing at the end of the dry season and in the first months of the rainy season. This observed relationship between season and the occurrence of preeclampsia raises new questions regarding the pathophysiology of preeclampsia. Possible explanations could be the impact of humidity and temperature on production of vasoactive substances. Dry and rainy seasons, through their influence on agricultural yields, may also impact on the nutritional status and play a role in the pathophysiology of the women.

Windstorms and Tropical Cyclones

Direct Consequences

In the 1991 cyclone disasters that killed 140,000 people in Bangladesh, 90% of victims were women. The death rate among people aged 20–44 years was 71 per 1000 women, compared with 15 per 1000 men. Explanations for this include the fact that more women than men are homebound, looking after children and valuables. Even if a warning is issued, many women die while waiting for their relatives to return home to accompany them to a safe place. Other reasons include the sari restricts the movement of women and puts them more at risk at the time of a tidal surge, and that women are less well-nourished and hence physically less able than men to deal with these situations. In May 2008, Cyclone Nargis came ashore in the Ayeyarwady Division of Myanmar. Among the 130 000 people dead or missing in the aftermath, 61% were female.

Indirect Consequences

Women, young people, and people with low socioeconomic status are thought to be at comparatively high risk of anxiety and mood disorders after disasters. One study of anxiety and mood disorder (as defined by the fourth edition of the Diagnostic and Statistical Manual of Mental Disorders; DSM-IV) after Hurricane Katrina found the incidence was consistently associated with the following factors: age under 60 years; being a woman; education level lower than college completion; low family income; pre-hurricane employment status (largely unemployed and disabled); and being unmarried. In addition, Hispanic people and people of other racial/ethnic minorities (not including non-Hispanic black people) had a significantly lower estimated incidence of any disorder compared with non-Hispanic white people in the New Orleans area, as well as a significantly lower estimated prevalence of posttraumatic stress disorder (PTSD) in the remainder of the sample. These same associations have been found in community epidemiological surveys in the absence of disasters, suggesting that these associations might be related to pre-existing conditions. A follow-up study that looked at patterns and correlates of recovery from hurricane-related PTSD, broader anxiety and mood disorders and suicidal behaviour found a high prevalence of hurricane related mental illness widely distributed in the population nearly 2 years after the hurricane.

Sea-Level Rises, Heavy Rain and Flooding

Increasing temperatures are contributing to sea-level rises, and precipitation is becoming heavier and more variable in many regions, potentially increasing flood risks and multiple associated health hazards. There has, however, been only limited systematic research and gender analysis on the health outcomes of flooding. It is important to recognize that vulnerability to flooding is differentiated by social dimensions. In both developing and industrialized nations, health and other impacts may fall disproportionately on women, children, people with disabilities and elderly people.

Direct Consequences

A report on the health effects of climate change in the United Kingdom showed that age- and gender-related information on flood deaths is incomplete. Published reviews have shown, however, that men are much more at risk of drowning than women, probably due to taking more risky or "heroic" behavior.

Saline contamination is expected to be aggravated by climate change and sea-level rises. A paper on saline contamination of drinking water in Bangladesh indicated that large numbers of pregnant women in coastal areas are being diagnosed with preeclampsia, eclampsia and hypertension. Although local doctors and community representatives have blamed the problem on increased salinity, no formal epidemiological study has been done.

Global Warming

Global warming is the long-term warming of the planet's overall temperature. Though this warming trend has been going on for a long time, its pace has significantly increased in the last hundred years due to the burning of fossil fuels. As the human population has increased, so has the volume of fossil fuels burned? Fossil fuels include coal, oil, and natural gas, and burning them causes what is known as the "greenhouse effect" in Earth's atmosphere.

The greenhouse effect is when the Sun's rays penetrate the atmosphere, but when that heat is reflected off the surface cannot escape back into space. Gases produced by the burning of fossil fuels prevent the heat from leaving the atmosphere. These greenhouse gasses are carbon dioxide, chlorofluorocarbons, water vapor, methane, and nitrous oxide. The excess heat in the atmosphere has caused the average global temperature to rise overtime, otherwise known as global warming.

Global warming has presented another issue called climate change. Sometimes these phrases are used interchangeably, however, they are different. Climate change refers to changes in weather patterns and growing seasons around the world. It also refers to sea level rise caused by the expansion of warmer seas and melting ice sheets and glaciers. Global warming causes climate change, which poses a serious threat to life on earth in the forms of widespread flooding and extreme weather.

Global warming is a serious issue and is not a single issue but a number of environmental issues. Global warming is a rise in the surface temperature of the earth that has changed various life forms on the earth. The issues that cause global warming is divided into two categories include "natural" and "human influences" of global warming.

Natural Causes of Global Warming

The climate has continuously changing for centuries. The global warming happens because the natural rotation of the sun that changes the intensity of sunlight and moving closer to the earth.

Another cause of global warming is greenhouse gases. Greenhouse gases are carbon monoxide and sulphur dioxide it trap the solar heats rays and prevent it from escaping from the surface of the earth. This has cause the temperature of the earth increase.

Volcanic eruptions are another issue that causes global warming. For instance, a single volcanic eruption will release amount of carbon dioxide and ash to the atmosphere. Once carbon dioxide increase, the temperature of earth increase and greenhouse trap the solar radiations in the earth.

Finally, methane is another issue that causes global warming. Methane is also a greenhouse gas. Methane is more effective in trapping heat in the atmosphere that carbon dioxide by 20 times. Usually methane gas can release from many areas. For instance, it can be from cattle, landfill, natural gas, petroleum systems, coal mining, mobile explosion, or industrial waste process.

Human Influences on Global Warming

Human influence has been a very serious issue now because human do not take care the earth. Human that cause global warming are more than natural causes global warming. The earth has been changing for many years until now it is still changing because of modern lifestyle of human. Human activities include industrial production, burning fossil fuel, mining, cattle rearing or deforestation.

First issue is industrial revolution. Industrial have been using fossil fuels for power machines. Everything that we use is involved in fossil fuel. For example, when we buy a mobile phone, the process of making mobile phone have involve machines and machines uses fossil fuels, during the process carbon dioxide is releasing to the atmosphere. Besides industrial, transportation such as cars is also releasing carbon dioxide from exhaust.

Another issue is mining. During the process of mining, the methane will trap below the earth. Besides, rearing cattle will also cause methane because cattle released the form of manure. However, cattle are important because it make the latter equally responsible for the occurrence of global warming.

Next is the most common issue that is deforestation. Deforestation is a human influence because human have been cutting down trees to produce papers, wood, build houses or more. If human continuing deforestation, carbon dioxide will concentrate in the atmosphere because trees can absorb carbon dioxide from atmosphere. Besides, human also release carbon dioxide when breathe. Therefore the amounts of millions of people breath have release carbon dioxide to the atmosphere. If human continue deforestation, human breathing that release carbon dioxide will stay at the atmosphere.

The Effect of Global Warming

The greenhouse gases will stay in the atmosphere for many years since hundreds years ago. However, the effect that global warming will cause on earth are extremely serious. There are many effects that will happen in the future if global warming continues. That includes polar ice caps melting, economic consequences, warmer waters and more hurricanes, spread of diseases and earthquake.

First effect is polar ice caps melting. As the temperature increase, the ice at the North Pole will melt. Once the ice melt the first effect will be raise on sea levels because the melting glaciers become oceans. According to the National Snow and Ice Data Center "if the ice melted today the seas would rise about 230 feet". It affects many low lying areas such as the Netherlands. In future, the Netherlands will be cover by water once the North Pole is melted. However, it is not going to happen so fast but the sea level will continue rise.

Another effect is the species loss of habitat. Species that include polar bears and tropical frogs will be extinct due to climate change. Besides, various birds will migrate to other places because animals are not like humans. They cannot adapt the habitat that changes their living or temperature.

Next effect is more hurricanes will occur and economic consequences still affect as well. Hurricane causes damage to houses and government need to spend billions of dollars in damage and people need places to stay or have been killed. Once a disaster happens many people have died and diseases happen. Diseases are more serious because it can spread to other people very fast and more people will get the disease and the disease maybe come more serious because of different weather.

Solution to Stop Global Warming

To reduce global warming we can start by reducing the use gasoline, wastage of electricity and our activities which are leading to global warming. To reduce gasoline mean we have a choice to choose a hybrid car that reduce using gasoline. Besides, petrol price are increasing. If a person everyday drives to work they need to pump petrol after 3 days and causes carbon dioxide. Another way to reduce gasoline is take public transport or carpool to work. It can help reduce carbon dioxide and save cost.

Another way to reduce global warming is recycle. Recycle can reduce garbage by reusing plastic bags, bottles, papers or glass. For instance, when we buy foods, we can use our own containers instead of plastic bags. Another example is after finish drinking the water from the bottle; we can reuse it or use our own bottle. If all this is being reuse, human can reduce deforestation and help save environment. Besides, turn off electricity if unused. It can save thousands of carbon dioxide and buy product that have energy saving because it saves cost and save environment.

Finally, human should stop open burning such as burning dry leafs or burning garbage. It will release carbon dioxide and toxic if burning garbage with plastic. Besides, government should reduce deforestation because the earth temperatures are increasing. Trees will help to improve the temperature on earth.

Chapter 6

Climate and its Models

A climate model can be essentially defined as a representation of the climate system based on physical, biological and chemical principles. Climate models help in understanding the complex Earth systems and draw conclusions on past and future climate systems. This chapter has been carefully written to provide the reader with a better understanding of climate models.

Climate Model

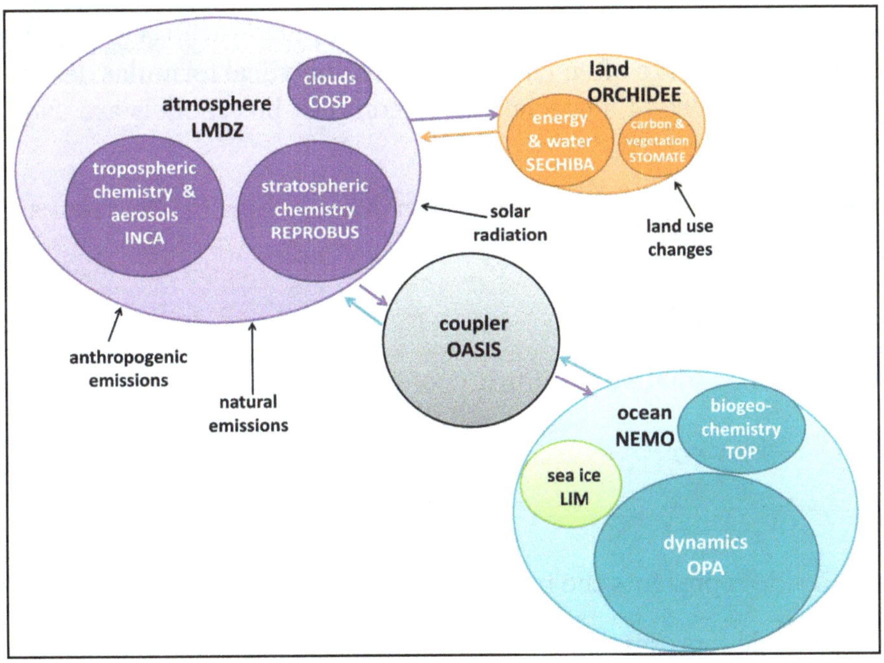

Climate models use quantitative methods to simulate the interactions of the important drivers of climate, including atmosphere, oceans, land surface and ice. They are used for a variety of purposes from study of the dynamics of the climate system to projections of future climate.

All climate models take account of incoming energy from the sun as short wave electromagnetic radiation, chiefly visible and short-wave (near) infrared, as well as outgoing long wave (far) infrared electromagnetic. Any imbalance results in a change in temperature.

Models vary in complexity:

- A simple radiant heat transfer model treats the earth as a single point and averages outgoing energy.

- This can be expanded vertically (radiative-convective models) and/or horizontally.
- Finally, (coupled) atmosphere–ocean–sea ice global climate models solve the full equa-tions for mass and energy transfer and radiant exchange.
- Box models can treat flows across and within ocean basins.
- Other types of modelling can be interlinked, such as land use, allowing researchers to pre-dict the interaction between climate and ecosystems.

Box Models

Box models are simplified versions of complex systems, reducing them to boxes (or reservoirs) linked by fluxes. The boxes are assumed to be mixed homogeneously. Within a given box, the concentration of any chemical species is therefore uniform. However, the abundance of a species within a given box may vary as a function of time due to the input to (or loss from) the box or due to the production, consumption or decay of this species within the box.

Simple box models, i.e. box model with a small number of boxes whose properties (e.g. their volume) do not change with time, are often useful to derive analytical formulas describing the dynamics and steady-state abundance of a species. More complex box models are usually solved using numerical techniques.

Box models are used extensively to model environmental systems or ecosystems and in studies of ocean circulation and the carbon cycle.

Zero-dimensional Models

A very simple model of the radiative equilibrium of the Earth is,

$$(1-a)S\pi r^2 = 4\pi r^2 \epsilon \sigma T^4$$

where

- The left hand side represents the incoming energy from the Sun.
- The right hand side represents the outgoing energy from the Earth, calculated from the Stefan-Boltzmann law assuming a model-fictive temperature, T, sometimes called the 'equilibrium temperature of the Earth', that is to be found,

and

- S is the solar constant – the incoming solar radiation per unit area—about 1367 $W{\cdot}m^{-2}$.
- a is the Earth's average albedo, measured to be 0.3.
- r is Earth's radius—approximately 6.371×10^6m.
- π is the mathematical constant (3.141...).
- σ is the Stefan-Boltzmann constant—approximately 5.67×10^{-8} $J{\cdot}K^{-4}{\cdot}m^{-2}{\cdot}s^{-1}$.

- ϵ is the effective emissivity of earth, about 0.612

The constant πr^2 can be factored out, giving,

$$(1-a)S = 4\epsilon\sigma T^4$$

Solving for the temperature,

$$T = \sqrt[4]{\frac{(1-a)S}{4\epsilon\sigma}}$$

This yields an apparent effective average earth temperature of 288 K (15 °C; 59 °F). This is because the above equation represents the effective *radiative* temperature of the Earth (including the clouds and atmosphere). The use of effective emissivity and albedo account for the greenhouse effect.

This very simple model is quite instructive, and the only model that could fit on a page. For example, it easily determines the effect on average earth temperature of changes in solar constant or change of albedo or effective earth emissivity.

The average emissivity of the earth is readily estimated from available data. The emissivities of terrestrial surfaces are all in the range of 0.96 to 0.99 (except for some small desert areas which may be as low as 0.7). Clouds, however, which cover about half of the earth's surface, have an average emissivity of about 0.5 (which must be reduced by the fourth power of the ratio of cloud absolute temperature to average earth absolute temperature) and an average cloud temperature of about 258 K (−15 °C; 5 °F). Taking all this properly into account results in an effective earth emissivity of about 0.64 (earth average temperature 285 K (12 °C; 53 °F)).

This simple model readily determines the effect of changes in solar output or change of earth albedo or effective earth emissivity on average earth temperature. It says nothing, however about what might cause these things to change. Zero-dimensional models do not address the temperature distribution on the earth or the factors that move energy about the earth.

Radiative-convective Models

The zero-dimensional model above, using the solar constant and given average earth temperature, determines the effective earth emissivity of long wave radiation emitted to space. This can be refined in the vertical to a one-dimensional radiative-convective model, which considers two processes of energy transport:

- Upwelling and downwelling radiative transfer through atmospheric layers that both absorb and emit infrared radiation.
- Upward transport of heat by convection (especially important in the lower troposphere).

The radiative-convective models have advantages over the simple model: they can determine the effects of varying greenhouse gas concentrations on effective emissivity and therefore the surface temperature. But added parameters are needed to determine local emissivity and albedo and address the factors that move energy about the earth.

Effect of ice-albedo feedback on global sensitivity in a one-dimensional radiative-convective climate model.

Higher-dimension Models

The zero-dimensional model may be expanded to consider the energy transported horizontally in the atmosphere. This kind of model may well be zonally averaged. This model has the advantage of allowing a rational dependence of local albedo and emissivity on temperature – the poles can be allowed to be icy and the equator warm – but the lack of true dynamics means that horizontal transports have to be specified.

Emics (Earth-system Models of Intermediate Complexity)

Depending on the nature of questions asked and the pertinent time scales, there are, on the one extreme, conceptual, more inductive models, and, on the other extreme, general circulation models operating at the highest spatial and temporal resolution currently feasible. Models of intermediate complexity bridge the gap. One example is the Climber-3 model. Its atmosphere is a 2.5-dimensional statistical-dynamical model with 7.5° × 22.5° resolution and time step of half a day; the ocean is MOM-3 (Modular Ocean Model) with a 3.75° × 3.75° grid and 24 vertical levels.

Gcms (Global Climate Models or General Circulation Models)

General Circulation Models (GCMs) discretise the equations for fluid motion and energy transfer and integrate these over time. Unlike simpler models, GCMs divide the atmosphere and/or oceans into grids of discrete "cells", which represent computational units. Unlike simpler models which make mixing assumptions, processes internal to a cell—such as convection—that occur on scales too small to be resolved directly are parameterised at the cell level, while other functions govern the interface between cells.

Atmospheric GCMs (AGCMs) model the atmosphere and impose sea surface temperatures as boundary conditions. Coupled atmosphere-ocean GCMs (AOGCMs, e.g. HadCM3, EdGCM, GFDL CM2.X, ARPEGE-Climat) combine the two models. The first general circulation climate model that combined both oceanic and atmospheric processes was developed in the late 1960s at the NOAA Geophysical Fluid Dynamics Laboratory AOGCMs represent the pinnacle of complexity in climate models and internalise as many processes as possible. However, they are still under development and uncertainties remain. They may be coupled to models of other processes, such as the carbon cycle, so as to better model feedback effects. Such integrated multi-system models are sometimes referred to as either "earth system models" or "global climate models."

Research and Development

There are three major types of institution where climate models are developed, implemented and used:

- National meteorological services. Most national weather services have a climatology section.
- Universities. Relevant departments include atmospheric sciences, meteorology, climatology, and geography.

- National and international research laboratories. Examples include the National Center for Atmospheric Research (NCAR, in Boulder, Colorado, USA), the Geophysical Fluid Dynamics Laboratory (GFDL, in Princeton, New Jersey, USA), the Hadley Centre for Climate Prediction and Research (in Exeter, UK), the Max Planck Institute for Meteorology in Hamburg, Germany, or the Laboratoire des Sciences du Climat et de l'Environnement (LSCE), France, to name but a few.

The World Climate Research Programme (WCRP), hosted by the World Meteorological Organization (WMO), coordinates research activities on climate modelling worldwide.

A 2012 U.S. National Research Council report discussed how the large and diverse U.S. climate modeling enterprise could evolve to become more unified. Efficiencies could be gained by developing a common software infrastructure shared by all U.S. climate researchers, and holding an annual climate modeling forum, the report found.

Climate Models on the Web

- Dapper/DChart — plot and download model data referenced by the Fourth Assessment Report (AR4) of the Intergovernmental Panel on Climate Change. (No longer available).
- NCAR/UCAR Community Climate System Model (CCSM).
- Do it yourself climate prediction.
- Primary research GCM developed by NASA/GISS (Goddard Institute for Space Studies).
- Original NASA/GISS global climate model (GCM) with a user-friendly interface for PCs and Macs.
- CCCma model info and interface to retrieve model data.
- NOAA/Geophysical Fluid Dynamics Laboratory CM2 global climate model info and model output data files.
- University of Victoria Global climate model, free for download. Leading researcher was a contributing author to an IPCC report on climate change.
- vimeo.com/user12523377/videos Visualizations of climate models of ETH Zurich.

General Circulation Model

A general circulation model (GCM) is a type of climate model. It employs a mathematical model of the general circulation of a planetary atmosphere or ocean. It uses the Navier–Stokes equations on a rotating sphere with thermodynamic terms for various energy sources (radiation, latent heat).

These equations are the basis for computer programs used to simulate the Earth's atmosphere or oceans. Atmospheric and oceanic GCMs (AGCM and OGCM) are key components along with sea ice and land-surface components.

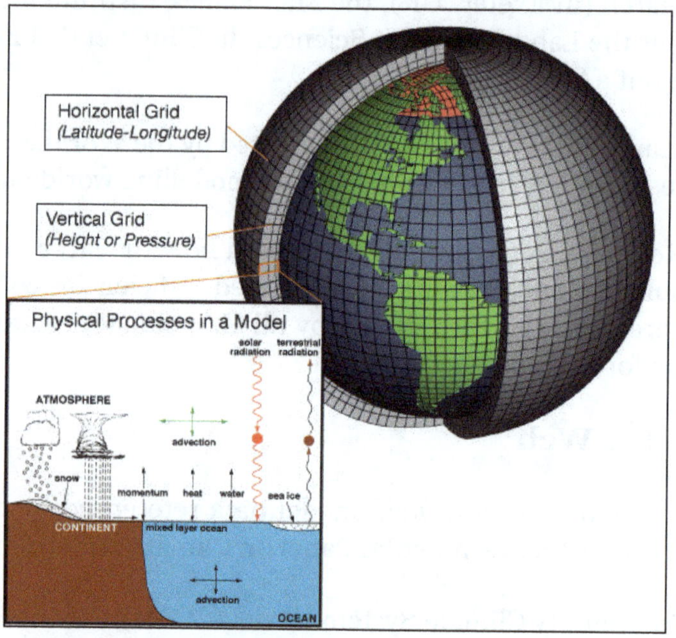

Climate models are systems of differential equations based on the basic laws of physics, fluid motion, and chemistry. To "run" a model, scientists divide the planet into a 3-dimensional grid, apply the basic equations, and evaluate the results. Atmospheric models calculate winds, heat transfer, radiation, relative humidity, and surface hydrology within each grid and evaluate interactions with neighboring points.

This visualization shows early test renderings of a global computational model of Earth's atmosphere based on data from NASA's Goddard Earth Observing System Model, Version 5 (GEOS-5).

GCMs and global climate models are used for weather forecasting, understanding the climate and forecasting climate change.

Versions designed for decade to century time scale climate applications were originally created by Syukuro Manabe and Kirk Bryan at the Geophysical Fluid Dynamics Laboratory in Princeton, New Jersey. These models are based on the integration of a variety of fluid dynamical, chemical and sometimes biological equations.

Terminology

The acronym *GCM* originally stood for *General Circulation Model*. Recently, a second meaning came into use, namely *Global Climate Model*. While these do not refer to the same thing, General Circulation Models are typically the tools used for modelling climate, and hence the two terms are sometimes used interchangeably. However, the term "global climate model" is ambiguous and may refer to an integrated framework that incorporates multiple components including a general circulation model, or may refer to the general class of climate models that use a variety of means to represent the climate mathematically.

History

In 1956, Norman Phillips developed a mathematical model that could realistically depict monthly and seasonal patterns in the troposphere. It became the first successful climate model. Following Phillips's work, several groups began working to create GCMs. The first to combine both oceanic and atmospheric processes was developed in the late 1960s at the NOAA Geophysical Fluid Dynamics Laboratory. By the early 1980s, the United States' National Center for Atmospheric Research had developed the Community Atmosphere Model; this model has been continuously refined. In 1996, efforts began to model soil and vegetation types. Later the Hadley Centre for Climate Prediction and Research's HadCM3 model coupled ocean-atmosphere elements. The role of gravity waves was added in the mid-1980s. Gravity waves are required to simulate regional and global scale circulations accurately.

Atmospheric and Oceanic Models

Atmospheric (AGCMs) and oceanic GCMs (OGCMs) can be coupled to form an atmosphere-ocean coupled general circulation model (CGCM or AOGCM). With the addition of submodels such as a sea ice model or a model for evapotranspiration over land, AOGCMs become the basis for a full climate model.

Trends

A recent trend in GCMs is to apply them as components of Earth system models, e.g. by coupling ice sheet models for the dynamics of the Greenland and Antarctic ice sheets, and one or more chemical transport models (CTMs) for species important to climate. Thus a carbon CTM may allow a GCM to better predict anthropogenic changes in carbon dioxide concentrations. In addition, this approach allows accounting for inter-system feedback: e.g. chemistry-climate models allow the possible effects of climate change on ozone hole to be studied.

Climate prediction uncertainties depend on uncertainties in chemical, physical and social models. Significant uncertainties and unknowns remain, especially regarding the future course of human population, industry and technology.

Structure

Three-dimensional (more properly four-dimensional) GCMs apply discrete equations for fluid motion and integrate these forward in time. They contain parameterisations for processes such as convection that occur on scales too small to be resolved directly.

A simple general circulation model (SGCM) consists of a dynamic core that relates properties such as temperature to others such as pressure and velocity. Examples are programs that solve the primitive equations, given energy input and energy dissipation in the form of scale-dependent friction, so that atmospheric waves with the highest wavenumbers are most attenuated. Such models may be used to study atmospheric processes, but are not suitable for climate projections.

Atmospheric GCMs (AGCMs) model the atmosphere (and typically contain a land-surface model as well) using imposed sea surface temperatures (SSTs). They may include atmospheric chemistry.

AGCMs consist of a dynamical core which integrates the equations of fluid motion, typically for:

- Surface pressure.
- Horizontal components of velocity in layers.
- Temperature and water vapor in layers.
- Radiation, split into solar/short wave and terrestrial/infra-red/long wave.
- Parameters for:
    - Convection,
    - Land surface processes,
    - Albedo,
    - Hydrology,
    - Cloud cover.

A GCM contains prognostic equations that are a function of time (typically winds, temperature, moisture, and surface pressure) together with diagnostic equations that are evaluated from them for a specific time period. As an example, pressure at any height can be diagnosed by applying the hydrostatic equation to the predicted surface pressure and the predicted values of temperature between the surface and the height of interest. Pressure is used to compute the pressure gradient force in the time-dependent equation for the winds.

OGCMs model the ocean (with fluxes from the atmosphere imposed) and may contain a sea ice model. For example, the standard resolution of HadOM3 is 1.25 degrees in latitude and longitude, with 20 vertical levels, leading to approximately 1,500,000 variables.

AOGCMs (e.g. HadCM3, GFDL CM2.X) combine the two submodels. They remove the need to specify fluxes across the interface of the ocean surface. These models are the basis for model predictions of future climate, such as are discussed by the IPCC. AOGCMs internalise as many processes as possible. They have been used to provide predictions at a regional scale. While the simpler models are generally susceptible to analysis and their results are easier to understand, AOGCMs may be nearly as hard to analyse as the climate itself.

Grid

The fluid equations for AGCMs are made discrete using either the finite difference method or the spectral method. For finite differences, a grid is imposed on the atmosphere. The simplest grid uses constant angular grid spacing (i.e., a latitude / longitude grid). However, non-rectangular grids (e.g., icosahedral) and grids of variable resolution are more often used. The LMDz model can be arranged to give high resolution over any given section of the planet. HadGEM1 (and other ocean models) use an ocean grid with higher resolution in the tropics to help resolve processes believed to be important for the El Niño Southern Oscillation (ENSO). Spectral models generally use a gaussian grid, because of the mathematics of transformation between spectral and grid-point space. Typical AGCM resolutions are between 1 and 5 degrees in latitude or longitude: HadCM3, for example, uses 3.75 in longitude and 2.5 degrees in latitude, giving a grid of 96 by 73 points (96 x 72 for some variables); and has 19 vertical levels. This results in approximately 500,000 "basic" variables, since each grid point has four variables (u,v, T, Q), though a full count would give more (clouds; soil levels). HadGEM1 uses a grid of 1.875 degrees in longitude and 1.25 in latitude in the atmosphere; HiGEM, a high-resolution variant, uses 1.25 x 0.83 degrees respectively. These resolutions are lower than is typically used for weather forecasting. Ocean resolutions tend to be higher, for example HadCM3 has 6 ocean grid points per atmospheric grid point in the horizontal.

For a standard finite difference model, uniform gridlines converge towards the poles. This would lead to computational instabilities and so the model variables must be filtered along lines of latitude close to the poles. Ocean models suffer from this problem too, unless a rotated grid is used in which the North Pole is shifted onto a nearby landmass. Spectral models do not suffer from this problem. Some experiments use geodesic grids and icosahedral grids, which (being more uniform) do not have pole-problems. Another approach to solving the grid spacing problem is to deform a Cartesian cube such that it covers the surface of a sphere.

Flux Buffering

Some early versions of AOGCMs required an *ad hoc* process of "flux correction" to achieve a stable climate. This resulted from separately prepared ocean and atmospheric models that each used an implicit flux from the other component different than that component could produce. Such a model failed to match observations. However, if the fluxes were 'corrected', the factors that led to these unrealistic fluxes might be unrecognised, which could affect model sensitivity. As a result, the vast majority of models used in the current round of IPCC reports do not use them. The model improvements that now make flux corrections unnecessary include improved ocean physics, improved resolution in both atmosphere and ocean, and more physically consistent coupling between atmosphere and ocean submodels. Improved models now maintain stable, multi-century simulations of surface climate that are considered to be of sufficient quality to allow their use for climate projections.

Convection

Moist convection releases latent heat and is important to the Earth's energy budget. Convection occurs on too small a scale to be resolved by climate models, and hence it must be handled via parameters. This has been done since the 1950s. Akio Arakawa did much of the early work, and variants of his scheme are still used, although a variety of different schemes are now in use. Clouds are also typically handled with a parameter, for a similar lack of scale. Limited understanding of clouds has limited the success of this strategy, but not due to some inherent shortcoming of the method.

Software

Most models include software to diagnose a wide range of variables for comparison with observations or study of atmospheric processes. An example is the 1.5-metre temperature, which is the standard height for near-surface observations of air temperature. This temperature is not directly predicted from the model but is deduced from surface and lowest-model-layer temperatures. Other software is used for creating plots and animations.

Projections

Projected annual mean surface air temperature from 1970-2100, based on SRES emissions scenario A1B, using the NOAA GFDL CM2.1 climate model (credit: NOAA Geophysical Fluid Dynamics Laboratory).

Coupled AOGCMs use transient climate simulations to project/predict climate changes under various scenarios. These can be idealised scenarios (most commonly, CO_2 emissions increasing at 1%/yr) or based on recent history (usually the "IS92a" or more recently the SRES scenarios). Which scenarios are most realistic remains uncertain.

The 2001 IPCC Third Assessment Report shows the global mean response of 19 different coupled models to an idealised experiment in which emissions increased at 1% per year. It shows the response of a smaller number of models to more recent trends. For the 7 climate models shown there, the temperature change to 2100 varies from 2 to 4.5 °C with a median of about 3 °C.

Future scenarios do not include unknown events – for example, volcanic eruptions or changes in solar forcing. These effects are believed to be small in comparison to greenhouse gas (GHG) forcing in the long term, but large volcanic eruptions, for example, can exert a substantial temporary cooling effect.

Human GHG emissions are a model input, although it is possible to include an economic/technological submodel to provide these as well. Atmospheric GHG levels are usually supplied as an input, though it is possible to include a carbon cycle model that reflects vegetation and oceanic processes to calculate such levels.

Emissions Scenarios

Projected change in annual mean surface air temperature from the late 20th century to the middle 21st century, based on SRES emissions scenario A1B (credit: NOAA Geophysical Fluid Dynamics Laboratory).

For the six SRES marker scenarios, IPCC (2007) gave a "best estimate" of global mean temperature increase (2090–2099 relative to the period 1980–1999) of 1.8 °C to 4.0 °C. Over the same time period, the "likely" range (greater than 66% probability, based on expert judgement) for these scenarios was for a global mean temperature increase of 1.1 to 6.4 °C.

In 2008 a study made climate projections using several emission scenarios. In a scenario where global emissions start to decrease by 2010 and then declined at a sustained rate of 3% per year, the likely global average temperature increase was predicted to be 1.7 °C above pre-industrial levels by 2050, rising to around 2 °C by 2100. In a projection designed to simulate a future where no efforts are made to reduce global emissions, the likely rise in global average temperature was predicted to be 5.5 °C by 2100. A rise as high as 7 °C was thought possible, although less likely.

Another no-reduction scenario resulted in a median warming over land (2090–99 relative to the period 1980–99) of 5.1 °C. Under the same emissions scenario but with a different model, the predicted median warming was 4.1 °C.

Model Accuracy

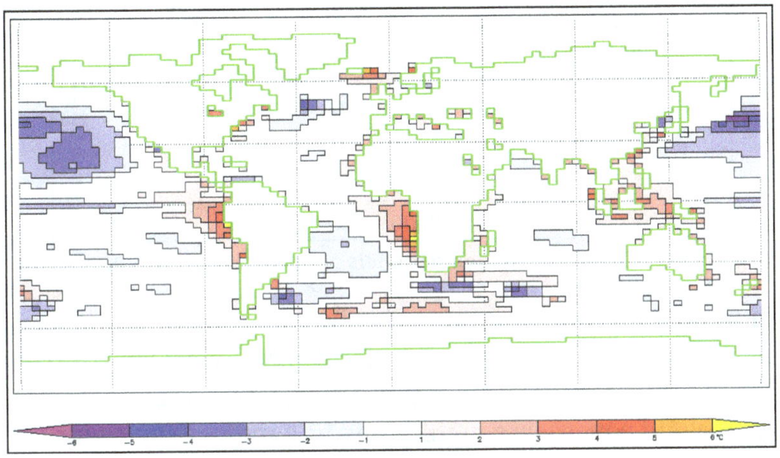

SST errors in HadCM3.

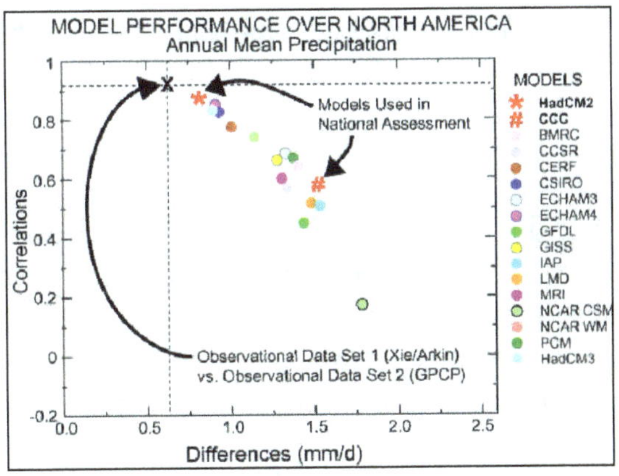

North American precipitation from various models.

AOGCMs internalise as many processes as are sufficiently understood. However, they are still under development and significant uncertainties remain. They may be coupled to models of other processes, such as the carbon cycle, so as to better model feedbacks. Most recent simulations show "plausible" agreement with the measured temperature anomalies over the past 150 years, when driven by observed changes in greenhouse gases and aerosols. Agreement improves by including both natural and anthropogenic forcings.

Imperfect models may nevertheless produce useful results. GCMs are capable of reproducing the general features of the observed global temperature over the past century.

A debate over how to reconcile climate model predictions that upper air (tropospheric) warming should be greater than observed surface warming, some of which appeared to show otherwise, was resolved in favour of the models, following data revisions.

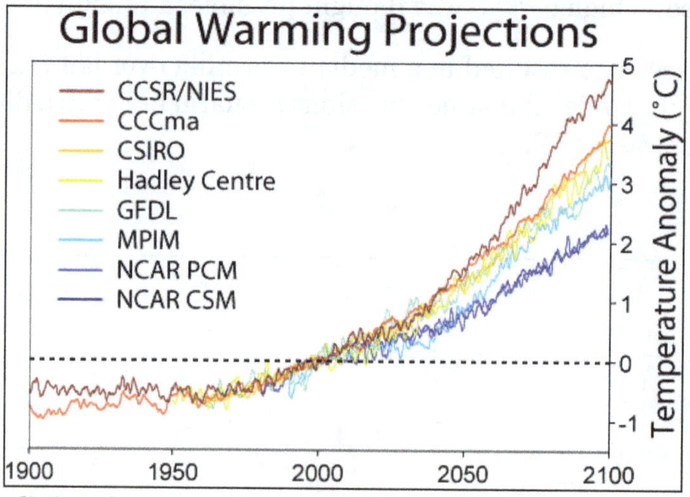

Temperature predictions from some climate models assuming the SRES A2 emissions scenario.

Cloud effects are a significant area of uncertainty in climate models. Clouds have competing effects on climate. They cool the surface by reflecting sunlight into space; they warm it by increasing the amount of infrared radiation transmitted from the atmosphere to the surface. In the 2001 IPCC report possible changes in cloud cover were highlighted as a major uncertainty in predicting climate.

Climate researchers around the world use climate models to understand the climate system. Thousands of papers have been published about model-based studies. Part of this research is to improve the models.

In 2000, a comparison between measurements and dozens of GCM simulations of ENSO-driven tropical precipitation, water vapor, temperature, and outgoing longwave radiation found similarity between measurements and simulation of most factors. However the simulated change in precipitation was about one-fourth less than what was observed. Errors in simulated precipitation imply errors in other processes, such as errors in the evaporation rate that provides moisture to create precipitation. The other possibility is that the satellite-based measurements are in error. Either indicates progress is required in order to monitor and predict such changes.

A more complete discussion of climate models is provided in the IPCC's Third Assessment Report.

- The model mean exhibits good agreement with observations.
- The individual models often exhibit worse agreement with observations.
- Many of the non-flux adjusted models suffered from unrealistic climate drift up to about 1 °C/century in global mean surface temperature.
- The errors in model-mean surface air temperature rarely exceed 1 °C over the oceans and 5 °C over the continents; precipitation and sea level pressure errors are relatively greater but the magnitudes and patterns of these quantities are recognisably similar to observations.

- Surface air temperature is particularly well simulated, with nearly all models closely matching the observed magnitude of variance and exhibiting a correlation > 0.95 with the observations.
- Simulated variance of sea level pressure and precipitation is within ±25% of observed.
- All models have shortcomings in their simulations of the present day climate of the stratosphere, which might limit the accuracy of predictions of future climate change.
    - There is a tendency for the models to show a global mean cold bias at all levels.
    - There is a large scatter in the tropical temperatures.
    - The polar night jets in most models are inclined poleward with height, in noticeable contrast to an equatorward inclination of the observed jet.
    - There is a differing degree of separation in the models between the winter sub-tropical jet and the polar night jet.
- For nearly all models the r.m.s. error in zonal- and annual-mean surface air temperature is small compared with its natural variability.
    - There are problems in simulating natural seasonal variability. (2000)
        - In flux-adjusted models, seasonal variations are simulated to within 2 K of observed values over the oceans. The corresponding average over non-flux-adjusted models shows errors up to about 6 K in extensive ocean areas.
        - Near-surface land temperature errors are substantial in the average over flux-adjusted models, which systematically underestimates (by about 5 K) temperature in areas of elevated terrain. The corresponding average over non-flux-adjusted models forms a similar error pattern (with somewhat increased amplitude) over land.
        - In Southern Ocean mid-latitudes, the non-flux-adjusted models overestimate the magnitude of January-minus-July temperature differences by ~5 K due to an overestimate of summer (January) near-surface temperature. This error is common to five of the eight non-flux-adjusted models.
        - Over Northern Hemisphere mid-latitude land areas, zonal mean differences between July and January temperatures simulated by the non-flux-adjusted models show a greater spread (positive and negative) about observed values than results from the flux-adjusted models.
        - The ability of coupled GCMs to simulate a reasonable seasonal cycle is a necessary condition for confidence in their prediction of long-term climatic changes (such as global warming), but it is not a sufficient condition unless the seasonal cycle and long-term changes involve similar climatic processes.

- Coupled climate models do not simulate with reasonable accuracy clouds and some related hydrological processes (in particular those involving upper tropospheric humidity). Problems in the simulation of clouds and upper tropospheric humidity, remain worrisome because the associated processes account for most of the uncertainty in climate model simulations of anthropogenic change.

The precise magnitude of future changes in climate is still uncertain; for the end of the 21st century (2071 to 2100), for SRES scenario A2, the change of global average SAT change from AOGCMs compared with 1961 to 1990 is +3.0 °C (5.4 °F) and the range is +1.3 to +4.5 °C (+2.3 to 8.1 °F).

The IPCC's Fifth Assessment Report asserted "...very high confidence that models reproduce the general features of the global-scale annual mean surface temperature increase over the historical period." However, the report also observed that the rate of warming over the period 1998-2012 was lower than that predicted by 111 out of 114 Coupled Model Intercomparison Project climate models.

Relation to Weather Forecasting

The global climate models used for climate projections are similar in structure to (and often share computer code with) numerical models for weather prediction, but are nonetheless logically distinct.

Most weather forecasting is done on the basis of interpreting numerical model results. Since forecasts are short—typically a few days or a week—such models do not usually contain an ocean model but rely on imposed SSTs. They also require accurate initial conditions to begin the forecast—typically these are taken from the output of a previous forecast, blended with observations. Predictions must require only a few hours; but because they only cover a one-week the models can be run at higher resolution than in climate mode. Currently the ECMWF runs at 40 km (25 mi) resolution as opposed to the 100-to-200 km (62-to-124 mi) scale used by typical climate model runs. Often local models are run using global model results for boundary conditions, to achieve higher local resolution: for example, the Met Office runs a mesoscale model with an 11 km (6.8 mi) resolution covering the UK, and various agencies in the US employ models such as the NGM and NAM models. Like most global numerical weather prediction models such as the GFS, global climate models are often spectral models instead of grid models. Spectral models are often used for global models because some computations in modeling can be performed faster, thus reducing run times.

Computations

Climate models use quantitative methods to simulate the interactions of the atmosphere, oceans, land surface and ice.

All climate models take account of incoming energy as short wave electromagnetic radiation, chiefly visible and short-wave (near) infrared, as well as outgoing energy as long wave (far) infrared electromagnetic radiation from the earth. Any imbalance results in a change in temperature.

The most talked-about models of recent years relate temperature to emissions of greenhouse gases. These models project an upward trend in the surface temperature record, as well as a more rapid increase in temperature at higher altitudes.

Three (or more properly, four since time is also considered) dimensional GCM's discretise the equations for fluid motion and energy transfer and integrate these over time. They also contain parametrisations for processes such as convection that occur on scales too small to be resolved directly.

Atmospheric GCMs (AGCMs) model the atmosphere and impose sea surface temperatures as boundary conditions. Coupled atmosphere-ocean GCMs (AOGCMs, e.g. HadCM3, EdGCM, GFDL CM2.X, ARPEGE-Climat) combine the two models.

Models range in complexity:

- A simple radiant heat transfer model treats the earth as a single point and averages outgoing energy.
- This can be expanded vertically (radiative-convective models), or horizontally.
- Finally, (coupled) atmosphere–ocean–sea ice global climate models discretise and solve the full equations for mass and energy transfer and radiant exchange.
- Box models treat flows across and within ocean basins.

Other submodels can be interlinked, such as land use, allowing researchers to predict the interaction between climate and ecosystems.

Other Climate Models

Earth-system Models of Intermediate Complexity (Emics)

The Climber-3 model uses a 2.5-dimensional statistical-dynamical model with 7.5° × 22.5° resolution and time step of 1/2 a day. An oceanic submodel is MOM-3 (Modular Ocean Model) with a 3.75° × 3.75° grid and 24 vertical levels.

Radiative-convective Models (RCM)

One-dimensional, radiative-convective models were used to verify basic climate assumptions in the '80s and '90s.

Mesothermal

In climatology, the term mesothermal is used to refer to certain forms of climate found typically in the Earth's Temperate Zones. It has a moderate amount of heat, with winters not cold enough to sustain snow cover. Summers are warm within oceanic climate regimes, and hot within continental climate regimes.

Origin of Term

The term is derived from two Greek words meaning "having a moderate amount of heat." This can be misinterpreted, however, since the term is actually intended to describe only the temperature conditions that prevail during the winter months, rather than those for the year as a whole.

Under the broadest definition, all places with an average temperature in their coldest month that is colder than 18°C, but warmer than −3°C, are said to have a mesothermal climate. In some climate classification schemes, however, this is divided into two segments, with a coldest-month average of 6°C being the line of demarcation between them; then only those locations with a coldest-month temperature of between −3°C and 6°C are reckoned as mesothermal, the label "subtropical" being applied to areas where the average temperature in the coldest month ranges from 6°C to 18°C.

Observing the narrower definition articulated above, the mesothermal locations are those where the winters are too cold to allow year-round photosynthesis, but not cold enough to support a fixed period of continuous snow cover every year.

Range

In the USA, the northern boundary line between mesothermal and microthermal ranges is north of Juneau and Sitka at the Pacific Ocean. It goes sharply south to about 38N latitude in the Rockies, then eastward across the lower Midwest to the East Coast near Boston. The southern boundary line between mesothermal and megathermal (or tropical) is across south Florida just above Palm Beach.

Summer

Summers in these places may be hot (that is to say, having an average temperature in their warmest month of 22°C or above) or merely warm (with the warmest month averaging between 10°C and 22°C). The hot-summer, or continental, mesothermal climate is encountered exclusively in the Northern Hemisphere, in the landmass interiors of Asia and North America and along their east coasts, while the most frequently seen example of a warm to cool-summer mesothermal climate is the oceanic climates found along the west coasts of all of the world's continents, roughly equidistant between the geographical tropical and polar zones.

List of mesothermal cities and their summer temperatures:

- Hong Kong-hot summer
- Milan-hot summer
- New York City-hot summer
- Tokyo-hot summer
- London-warm summer
- Mexico City-warm summer
- Vancouver-warm summer

Moisture

In addition to being subdivisible by summer temperature, mesothermal climates can also be subclassified on the basis of precipitation — into humid, semiarid and arid subtypes within your front door within depth.

Microthermal

The word *microthermal* is derived from two Greek words meaning "small" and "heat". This is misleading, however, since the term is intended to describe only the temperature conditions that prevail during the winter months, rather than those of the entire year.

The characteristic feature of the microthermal climate is cold winters — specifically, winters that are cold enough to ensure that snow will remain on the ground continuously for a fixed period of time every year. Conceptually, an average temperature of 0°C or colder is assumed to be necessary to bring this about; thus the climate of a location where at least one full month is this cold is classified as microthermal (however, at least one month in the summer must average 10°C or higher; otherwise the climate would be reckoned as polar). This definition places all of the world's microthermal climates in the Northern Hemisphere, as the absence of broad land masses at upper-middle latitudes in the Southern Hemisphere precludes the existence of such temperature conditions there.

Microthermal climates are typically subdivided into three categories based on the temperature characteristics of the summer season. The southernmost of the three is frequently referred to as the temperate continental climate, and has hot summers — that is to say, at least one month has an average temperature of 22°C (71.6°F) or above. The middle zone is often labelled hemiboreal, and no summer month there has an average temperature as warm as 22°C, but at least four months will average 10°C (50°F) or higher. The northernmost of the three microthermal zones is the subarctic, or boreal zone; there only one to three months will have average temperatures of at least 10°C.

In North America, microthermal climates commence north of Boston along the Atlantic seaboard, this line drifting gradually southward further inland, reaching approximately 38° at the eastern edge of the Rocky Mountains, then curving dramatically northward near the Pacific coast, reaching the Pacific Ocean just south of Juneau, Alaska. In Asia, the latitude at which these climates begin is several degrees lower due to the pervasive influence of the vast Siberian anticyclone, or high-pressure system, and in continental Europe the line actually runs longitudinally rather than latitudinally, cutting through central Poland after beginning north of the Arctic Circle along the Norwegian coast, thereafter moving diagonally across Scandinavia.

The boundary between the microthermal and polar climate zones is farthest north in western Europe (actually within the Arctic Circle there), and farthest south along the east coast of North America (at about 56° North latitude on the central coast of Labrador); it then trends northward across Canada before dropping south again as it courses through Alaska. Throughout most of Siberia, the boundary tends to follow the Arctic Circle fairly closely.

In addition to having various summer temperature regimes, microthermal climates also differ from one another in how much precipitation they receive — such climates may be humid, semiarid or arid. Most of the Turkestan-Gobi desert system has an arid microthermal climate, while the best-known example of the semiarid microthermal climate can be found in the "steppes of Central Asia" immortalized by Russian classical music composer Alexander Borodin.

Megathermal

In climatology, the term megathermal (or less commonly, macrothermal) is sometimes used as a synonym for "tropical."

In order for a particular place to qualify as having a megathermal climate, every single month out of the year must have an average temperature of 18°C or above.

Megathermal climates are sometimes split into two temperature-based subsets — equatorial and tropical (the latter used here in the sense of "outer tropical") — with "equatorial" denoting little or no variation in temperature throughout the year and "tropical" denoting significant seasonal variation, even though no month has an average temperature of below 18°C. In addition, what temperature fluctuations do exist in an equatorial climate will typically bear no relationship to the astronomical seasons for the applicable side of the equator, while in the (outer) tropical subtype the temperature will move in concert with the seasons, the time of higher sun and longer days being warmest and the time of lower sun and shorter days coolest.

These climates can also be subdivided on the basis of rainfall, as examples of humid, semiarid and arid places can all be found within the megathermal category (although the region of arid megathermal climate is small compared with the extent of deserts existing in other climate zones, particularly the neighboring subtropical zone).

References

- "Pubs.GISS: Sun and Hansen 2003: Climate simulations for 1951-2050 with a coupled atmosphere-ocean model". pubs.giss.nasa.gov. 2003. Retrieved 04, August 2020
- "High Resolution Global Environmental Modelling (HiGEM) home page". Natural Environment Research Council and Met Office. 18 May 2004. Retrieved 11, May 2020
- Ken, Richard A (13 April 2001). "Global Warming: Rising Global Temperature, Rising Uncertainty". Science. 292 (5515): 192–194. doi:10.1126/science.292.5515.192. PMID 11305301. Retrieved 03, July 2020
- Lynch, Peter (2006). "The ENIAC Integrations". The Emergence of Numerical Weather Prediction. Cambridge University Press. pp. 206–208. ISBN 978-0-521-85729-1
- "Atmospheric Model Intercomparison Project". The Program for Climate Model Diagnosis and Intercomparison, Lawrence Livermore National Laboratory. Retrieved 20, January 2020
- C. Jablonowski, M. Herzog, J. E. Penner, R. C. Oehmke, Q. F. Stout, B. van Leer, January 1960.5091 "Adaptive Grids for Weather and Climate Models" (2004)

Permissions

All chapters in this book are published with permission under the Creative Commons Attribution Share Alike License or equivalent. Every chapter published in this book has been scrutinized by our experts. Their significance has been extensively debated. The topics covered herein carry significant information for a comprehensive understanding. They may even be implemented as practical applications or may be referred to as a beginning point for further studies.

We would like to thank the editorial team for lending their expertise to make the book truly unique. They have played a crucial role in the development of this book. Without their invaluable contributions this book wouldn't have been possible. They have made vital efforts to compile up to date information on the varied aspects of this subject to make this book a valuable addition to the collection of many professionals and students.

This book was conceptualized with the vision of imparting up-to-date and integrated information in this field. To ensure the same, a matchless editorial board was set up. Every individual on the board went through rigorous rounds of assessment to prove their worth. After which they invested a large part of their time researching and compiling the most relevant data for our readers.

The editorial board has been involved in producing this book since its inception. They have spent rigorous hours researching and exploring the diverse topics which have resulted in the successful publishing of this book. They have passed on their knowledge of decades through this book. To expedite this challenging task, the publisher supported the team at every step. A small team of assistant editors was also appointed to further simplify the editing procedure and attain best results for the readers.

Apart from the editorial board, the designing team has also invested a significant amount of their time in understanding the subject and creating the most relevant covers. They scrutinized every image to scout for the most suitable representation of the subject and create an appropriate cover for the book.

The publishing team has been an ardent support to the editorial, designing and production team. Their endless efforts to recruit the best for this project, has resulted in the accomplishment of this book. They are a veteran in the field of academics and their pool of knowledge is as vast as their experience in printing. Their expertise and guidance has proved useful at every step. Their uncompromising quality standards have made this book an exceptional effort. Their encouragement from time to time has been an inspiration for everyone.

The publisher and the editorial board hope that this book will prove to be a valuable piece of knowledge for students, practitioners and scholars across the globe.

Index

www.ingramcontent.com/pod-product-compliance
Lightning Source LLC
LaVergne TN
LVHW080723210926
848485LV00003B/512